As Though People Mattered

As Though People Mattered

A prospect for Britain

John Davis and Alan Bollard

With best wishes
John Davis

PRACTICAL ACTION
Publishing

Practical Action Publishing Ltd
The Schumacher Centre
Bourton on Dunsmore, Rugby,
Warwickshire CV23 9QZ, UK
www.practicalactionpublishing.org

Revised edition with new foreword and preface, 2013

ISBN 978-1-85339-809-4 Hardback
ISBN 978-1-85339-810-0 Paperback
ISBN 978-1-78044-809-1 Library Ebook
ISBN 978-1-78044-810-7 Ebook

Davis, J. and Bollard, A. (2013) *As Though People Mattered: A Prospect for
Britain*, Rugby, UK: Practical Action Publishing.

Since 1974, Practical Action Publishing (formerly Intermediate
Technology Publications and ITDG Publishing) has published and
disseminated books and information in support of international
development work throughout the world. Practical Action Publishing is
a trading name of Practical Action Publishing Ltd (Company Reg. No.
1159018), the wholly owned publishing company of Practical Action.
Practical Action Publishing trades only in support of its parent charity
objectives and any profits are covenanted back to Practical Action
(Charity Reg. No. 247257, Group VAT Registration No. 880 9924 76).

Typeset by Practical Action Publishing
Printed in the United Kingdom

Contents

Foreword

They may not quite be the four horsemen of the apocalypse, but there is no doubt that humanity's basic economic problem has four inter-linked parts.

First, we are already living well beyond our planet's capacity to regenerate itself. Many of our ecosystems are at risk of collapse, from fish stocks to coral reefs, and from fresh water to pollination systems. Above all, we face both peak oil and the real prospect of catastrophic climate change. Second, not only is global inequality in income and wealth untenable – the richest 1 per cent of people earn as much as the poorest 57 per cent – but trying to grow the world out of poverty by raising everybody's incomes further, without redistribution, is ecologically impossible. The earth just can't provide the resources that the growth would need. Third, our economic system is highly unstable. That was confirmed in the banking collapse. Fourth, for many people on earth, 'more' and 'better' have parted company. More wealth is not translating into greater well-being.

Each of these problems is recognized by policy-makers, but only to some degree. What they don't recognize is the way they link together, their systemic nature and their inter-relationships. That is why governments seem to be all at sea when it comes to solutions. What is on offer is a return to 'business as usual' with a green tint.

What is needed instead is an economic transformation, and this is why it is so exciting that *As Though People Mattered* is being republished. At the heart of such a transformation is a move from a 'consumer' to a 'conserver' society. This includes, as set out in the book, a focus on maintenance, repair, reconditioning, re-use and recycling. It involves thinking differently about productivity in many sectors – if our well-being depends on productive and good employment, as the New Economics Foundation's work has shown, and the most scarce resources are materials and energy, then we should be seeking to maximize employment per scarce unit of natural resource. We would need to give a fair return to capital, but break from the treadmill of providing ever-increasing returns to capital.

This would free us from the 'drug' of pursuing growth as the key goal of policy and relegate growth (or not) to a secondary outcome. The goals of our society would then be to increase well-being in ways that are socially just within fair environmental limits and to provide enough employment and public goods for all. This is totally feasible following the types of approaches set out in this book. The small business sector would thrive, many more people would be involved to some degree in food production, and finance would need to be transformed from a 'master' to a 'servant' of human endeavour.

This book brilliantly provides a clear view of what we need to change and how to do it.

Stewart Wallis
Executive Director, New Economics Foundation

Preface to the revised edition

The present financial and economic crisis is by no means the first time that Britain has been in a mess; but it is probably the most serious since World War II. On previous occasions we still had a reasonably balanced economy, with the private sector contributing about two-thirds of wealth generation, and imports and exports being roughly equal. Now the private sector contribution has decreased to little more than half, with government spending having outgrown its capacity to raise tax revenues by as much as £150 billion a year. At the same time in the private sector, as a result of deliberate government policies, the contribution from agriculture, fisheries and manufacturing has declined to such an extent that there is now a payment deficit of £20 billion on food and £55 billion on manufactured goods. Since 1980 the contribution from manufacturing has reduced from 25% to a mere 12% of gross domestic product. The theory has been that by concentrating on the growth of financial and business services these payment deficits would be affordable. In fact they have failed by as much as £40 billion. It has been a very costly experiment, leaving us with the massive twin tasks of rebalancing the public and private sectors in order to reduce the present record public sector deficit; and also to rebuild agriculture and manufacturing so that we can once again live within our means.

The two bits of wisdom that we need to re-learn are as follows:

> If you want to be a slave and pay the cost of your own slavery, then let the banks create money. [The present average personal debt excluding mortgages is £9,716.]
>
> Josiah Stamp, former director of the Bank of England

> I sympathize with those who would minimize, rather than those who would maximize, economic entanglements among nations. Ideas, knowledge, science, hospitality and travel – these are things that of their nature should be international. But let goods be home-spun wherever it is reasonable and conveniently possible; and above all, let finance be primarily national.
>
> John Maynard Keynes

But the present financial and economic crisis could not have happened at a more threatening time. Because we are already faced with two other life-changing challenges. They are an imminent oil supply crisis and climate change. Oil, which provides nearly all the energy required for transport and travel, is expected by most authorities to be unable to satisfy growing worldwide demand within the next 10 years. That will suddenly have a dramatic effect by increasing hugely all energy prices. In the long run that cannot be prevented because oil is a limited resource, as are the two other present sources of energy – coal and natural gas. But the onset of very high energy prices can be delayed, and become more affordable, if all energy use systems are quickly re-engineered to use only a fraction of present consumption. The potential threat posed by greatly increased energy prices was recognized 50 years ago. By 1980 it had become so widely recognized that the president of the European Economic Council (now the European Union) made the following extraordinary statement:

> Energy conservation must become the cornerstone of our policy. ... The potential saving is immense. It is now certain that, if we do not change our ways whilst there is still time, our society will risk dislocation and eventual collapse.

Sadly it has not been 'the cornerstone of our policy'; and there is now very little time left to make the necessary changes. Consequently it must now be given top priority in the rebalancing of the British economy in order to avoid the most serious consequences which the president referred to.

Unlike the suddenness of the oil crisis, climate change is a more gradual process. Nevertheless, if its dangerous long-term effects are to be avoided, urgent action is required to reduce British carbon dioxide emissions by more than 80 per cent in 2050 as our contribution to a worldwide response to the potential threat. That reduction can only be realized by a combination of more rational and efficient uses of all energy and the replacement of oil, coal and natural gas by alternative non-carbonaceous resources, such as renewables and nuclear.

Neglect of these two very dangerous threats over many years is, in some ways, similar to the neglect in the 1930s to re-arm in the face of the Nazi threat. It has become necessary to focus all our resources on reducing energy consumption and replacing fossil sources with renewables and nuclear sources; just as we concentrated effort on armaments and military in World War II in order to preserve a decent way of life.

Set alongside these two challenges, the most immediate concern arising in Britain, resulting from the neglect over recent years of farming and manufacturing, is the high and growing level of unemployment, as employment levels in the public sector can no longer be afforded. It was the highest level of unemployment since the 1930 slump which developed in the 1980s that led to a practical investigation of the potential for employment in small-scale private enterprise. It is a description of that investigation, and the lessons learned from it, that is the subject of this book. Although, over the past 25 years, a great deal of change has taken place in Britain, as has been described in the first paragraph of this preface, the experience of the early '80s is relevant to our present predicament and needs to be applied.

The important, and hopeful, difference is the immediate necessity to focus the development of all new enterprise on minimizing energy consumption and building the new low carbon energy supply system. This is where most of the new employment opportunities are to be found as investment is concentrated on this enormous task. The new enterprises and jobs will not disappear after a few years, as they did when armament production was run down at the end of the war. This time they will be permanent, providing all the work necessary to sustain the new energy use and supply systems for all our needs.

So even in the British private sector of the economy in which the 'rules of the game' favour big business and have done great damage to small and medium-sized firms, the prospects for regeneration of employment in SMEs looked hopeful 25 years ago. Those hopes have continued to be justified since then.

But what now are the prospects for employment in a completely different kind of society which, instead of year by year increasing consumption of materials and energy, seeks to conserve increasingly expensive resources. Let there be no doubt that we are not faced with a choice either to continue with the present system of 'consumerism' or change voluntarily to a 'conserver' system. We shall be forced to change to live within the limits which nature demands, in order to avoid further destruction of life and the means of sustaining life, or face the 'dislocation and eventual collapse' of which the president of the EEC spoke in 1980.

So we can begin to see 'conserver' society work as no longer increasingly in money earning as employees only, but with individuals beginning to take control of their working lives. Already there is an upsurge of people growing their own vegetables and fruit in their gardens, and a growing demand for allotments or use of spare patches of waste land by people without gardens. The example of Cuba, following its loss of imported oil, illustrates the potential that exists in towns and cities for individuals to grow a good proportion of their food. There is clearly xii AS THOUGH PEOPLE MATTERED

a growing urge amongst people, who find employment unsatisfying, to recover the enjoyment of working for themselves. Sometimes it is to earn money – between 1980 and 1990 self-employment increased from 2 million to 3.3 million, and it still continues beyond 4 million. Otherwise, it is simply to engage in constructive work, to substitute for produce, goods or services which instead would have to be purchased.

A picture of what a 'conserver' society would look like was painted by James Robertson in his valuable 1998 book *Beyond the Dependency Culture: People, Power and Responsibility*. He described it as follows:

- *Paid and unpaid work.* In manufacturing and services alike there will be further automation of large enterprises supplying mass-produced products and impersonal services. Many people will move into more personal work – on small farms, in small firms, in small community enterprises, and in the provision of goods and services on a person-to-person basis. There will be more people working in their own homes and neighbourhoods than there are today; more part-time work; and a fairer distribution of paid and unpaid work between men and women.
- *Industry.* There will be a continuing shift of emphasis towards the recovery and recycling of all kinds of materials, and methods of economizing in their use; a shift towards more durable goods, and therefore away from production towards servicing, maintenance and repair; and a shift towards the manufacture of small-scale technologies and equipment for small enterprises and do-it-yourself activities.
- *Food production and consumption.* Changes in agriculture and diet will make countries like Britain more self-sufficient in indigenous types of food and less dependent on imported produce and products. There will be less meat in the diet, more small farms, part-time farms, and do-it-yourself food-growing.
- *Patterns of settlement and patterns of living.* There will be a more dispersed pattern countrywide; more people living close to their work; more people growing food in cities; more people manufacturing in the country; and more people providing services directly to other people in both; more people spending more active time in and around their own home and locality; more people with their own food plots and workshops; increasing investment by households and neighbourhoods in many kinds of equipment to share, including mechanical, electrical tools and telecommunications equipment; more living and working together by children, young people, adults, and the elderly.

- *Decentralization and greater self-sufficiency.* In general, there will be a shift away from centralization towards greater autonomy and self-sufficiency at local and regional levels. In particular, localities and regions will strive to become less dependent on external sources of food and energy, recognizing that such dependency drains the local balance of payments and that local production for local consumption keeps money circulating locally, creating local jobs and a healthy local economy.

Unfortunately, many people wrongly believe that a high standard of living depends on a high level of consumption and high levels of waste. They do not understand that their domestic energy bills would be twice as expensive if the waste from 1970s boilers had not been halved by efficient modern boilers, and could be four times as expensive if their property had not been well insulated. Waste is already very impoverishing; and will become much more so as energy, food and all other resources become more expensive as a result of scarcity. The replacement of cheap fossil energy by alternatives, which almost all come as electricity, will be three times at least as expensive.

Each year we spend £300 billion to replace what has been imported, consumed – food/fuel – or thrown away after short periods of use – 2 tonnes per person per year! By doubling the period of usefulness of those imported goods and spending much of the money saved on life-extensive (labour-intensive) services, more than 1 million jobs, paying £25,000 a year, could be created and millions of tonnes of waste materials and energy saved. Furthermore, there would not be the continual pressure to increase exports to pay for ever increasing imports.

Although governments show no sign of abandoning 'business as usual', the creation of 'Transition Towns' is a hopeful sign that local community business initiatives could begin the process of conversion from a consumer society to a conserver society. This is the road to prosperity and fairness.

Preface to the first edition

In 1975 the Intermediate Technology Development Group responded to a growing demand for attention to be paid to the application of 'appropriate technology' concepts to Britain. To avoid distraction from its dedication to the Third World, John Davis was appointed to act on behalf of the Group on an 'AT – UK Project'. The intention was to contribute to 'economic development as though people mattered' by means of small scale enterprises which were sparing in their consumption of scarce resources, modest in their capital requirements, and able to provide satisfying employment.

Many obstacles stood in the way of such a development. These were partially responsible for the exceptional weakness of the British small businesses sector. (A recent report of the Economist Advisory Group which compares 'craft enterprises' in Britain and Germany estimates that 3.4 million are employed in such firms in Germany compared with only 1.4 million in Britain.) One of the most basic obstacles was a deeply rooted disbelief that such enterprises could be appropriate in modern Britain where big business ruled. Demonstration seemed to be the most effective way of overcoming this particular obstacle; and local community groups were the chosen medium for that demonstration. As a response to growing concern about unemployment, Local Enterprise Trusts began to be developed. They quickly showed that a significant potential for accelerated grass roots economic development existed in a variety of localities. Their effectiveness has encouraged such a widespread adoption of this type of community institution that there were almost 200 in operation throughout the UK in 1985.

Concern about the slow response in Britain to energy saving and material recycling later prompted the beginning of a parallel development of Local Energy Groups. These are dedicated to making their communities more efficient in their use of scarce resources.

Alongside these demonstrations of the value of 'economic development as though people mattered', a six-monthly newsletter was circulated to many influential people, and John Davis addressed a great many meetings and conferences at which theory and practice were discussed.

In 1980 Dr Alan Bollard joined the AT – UK Unit of ITDG to investigate the extent to which the balance between the big and small sectors of the UK economy might be shifted in favour of small scale. In addition to an examination of prospects in several different industrial sectors, he was able to identify the obstacles that tend to inhibit such a shift of emphasis. The results of his work have been presented in his book *Small Beginnings* (IT Publications, 1983).

As Though People Mattered has been written to bring together the main lessons that have been learned between 1975 and 1983 from all the work that has been done on this ITDG project. The contents have been chosen primarily because they were the particular aspects which, at meetings and conferences as well as in day-to-day operations, seem to be of greatest interest to the general public. It is a book aimed at a general readership rather than any specialist group.

Chapters 1, 2, 3, 4, 5, 9, and 10 have been written by John Davis and Chapters 6, 7, and 8 are the work of Alan Bollard. Each has commented on the other's contributions, but no attempt has been made to produce a uniform style. By dividing the work in this way advantage has been taken of the different professional expertise and experience of the two authors.

Chapter 2 may appear to some readers to be elementary and unnecessary. On very many occasions at public meetings there has been a great deal of confusion because of a lack of understanding of the interrelationships between the various elements involved in 'adding value'. Much of what is written in subsequent chapters cannot be understood unless there is a clear comprehension of these relationships.

Since the AT – UK Project was started in 1975 the significance of the small firm, a comparatively labour-intensive part of the economy, has come to be somewhat better appreciated in Britain and in similar advanced industrial countries. The importance of the careful steward-ship of scarce and nonrenewable resources is also gradually becoming more widely recognized as a necessary condition for a substantial global future.

It was not the late E.F. Schumacher's conviction that all that is big is bad, and all problems will evaporate if everything is done on a small scale. The authors of this book also do not wish to overstate the value of the small scale sector of the economy. It is, however, hoped that the experiences of the last ten years of the AT – UK Project may help to convince that there are alternative ways of dealing with the problems of production and distribution, involving a bigger role for small scale, resource economical enterprises, which are more human and satisfying.

Introduction

This book is about the beginning of a change in the direction of economic development in Britain. There is a widespread – even dominant – belief that there is no alternative to the form of development that has prevailed so far. It is widely believed that if we want electricity it has to be from a few giant power stations. If we want consumer durables they should not be built to last, but should be replaced after a few years' use. Most of our bread and beer must come from a small number of big bakeries and breweries operated by a handful of companies. It is better to make things rather than repair or recondition them. And so on.

As a consequence of this unswerving belief Britain has an industrial and commercial system that is more highly concentrated and centrally directed, in both private and public sectors, than that of any advanced Western nation. Compared with countries like France, West Germany, Italy, Sweden, and Denmark, little remains of local production meeting local needs. When comparisons are made about the use of scarce resources, Britain in many instances again compares very unfavourably in wastefulness.

Before we consider what alternative direction of development to follow, we review what has happened so far, what price has been paid to offset the advantages gained, how changes have come about, what expectations exist, and what is likely to be the consequence of adhering to our traditional system of development.

That examination makes clear that many of the disadvantages of the present system are inevitable byproducts of a single-minded commitment to 'consumerism'. It is this which tends to make modern society's institutions excessively unmanageable, hierarchical, bureaucratic, secretive, competitive, and wasteful. Within such a system it is very difficult for most people fully to participate and experience fulfilment. Man is set against man; man is diminished by machines and institutions; and man is made subservient to money. In a global context the very high level of consumption of increasingly scarce resources creates a widening gap between rich and poor. If it were to continue unchecked it could be disastrous.

If there is to be any hope of a just, participative, and sustainable future, one vitally important task is to discover ways in which a high standard and quality of life can be achieved on a smaller, more human scale, consuming only a fraction of scarce non-renewable resources. It means a gradual shift from big and wasteful towards small, efficient and frugal. It may be expressed as a transition from a 'consumerist' system to a 'conserver' system.

Such a radical change raises many worrying questions among people who fear that any departure from the present system, which has brought such obvious benefits to most people in the industrialized world, could be to throw out the baby with the bath-water. Also, the suggested change calls into question many firmly held convictions about the relationships between capital investment, productivity, economies of scale, employment, and technological development. These relationships are examined in Chapters 2-6; and succeeding chapters demonstrate that there is plenty of freedom and scope for a 'conserver' type approach to economic development. In the production sector most materials processing industries would remain large scale and employ little labour; while manufacturing of intermediate and finished products could mostly be carried out efficiently in small or medium-sized units. The important improvements in the use of scarce resources could partly be achieved by feasible changes in technology, but also by maintenance, repair, reconditioning, re-use, and recycling operations playing a much greater part in the economy.

The remaining chapters of the book deal with the practical ways in which a new thrust for a 'conserver' type development need to be, and are being, tackled in Britain. A big potential for small business development is described, together with the changes required in legislation to enable the full potential to be realized. Because this type of grass roots development necessarily requires a great deal of initiative and risk to be taken by many individuals there is need for new local institutions to be developed to act as catalytic promoters and provide community support for individual initiatives. Examples of local initiatives that have appeared in Britain in the past ten years, with increasing public and official support, are described. They are playing an important part, both directly in assisting new economic activities and also in building a new sense of community responsibility.

In the concluding chapter the narrow focus on Britain is widened to a global context. It points to the need for a similar 'conserver' type development to be adopted both in other industrialized countries and in the countries of the Third World. The achievement of that change in direction could halt the widening gap between rich and poor, and equally importantly prevent the environmental disasters that are widely forecast as the consequence of adherence to the existing 'consumerist' type of development.

CHAPTER 1

The changes afoot

How lives have changed as a result of innovations in technology and trends in society

Peoples' lives can be examined from many different points of view. Since this book is primarily concerned with the economic aspects of life in Britain, and particularly with the impact of changing technology, a convenient and simple method of discerning the changes that have occurred in people's lives is to consider three aspects – working life, domestic life, and family life and leisure – and what has happened in these areas within living memory – since, say, the First World War.

Working life

Between the two world wars remarkably little change in the way work was done resulted from the introduction of new production technologies. Even by 1939 tractors had by no means replaced horses on the land. Most crop harvesting was still carried out by traditional methods, and intensive animal rearing had not been introduced to any extent. Although in the materials producing industries great changes took place because of the Depression and intensifying foreign competition in export markets, particularly for steel and textiles, little investment was made in new plant and machinery of an advanced type. In engineering works methods changed very little. Although towards the end of the thirties electrically powered machines were coming into use, traditional shaft and belt drives to old machines were commonplace. Indeed, some of the lathes and milling machines used to produce Rolls Royce Merlin engines during the Battle of Britain were relics of the First World War, and automatic and semi-automatic machines were by no means universal in wartime factories. The main change that took place between the wars in manufacturing industry was the growth of new industries, which appeared mostly in the Midlands and the south of Britain, while the traditional heavy industries in the north – steel, coal, shipbuilding, heavy engineering, and textiles – began to decline. The three key elements of the new industries were electrical and electronic engineering, with its associated electrical appliances, transport – bicycles, cars, buses, and aircraft – and chemicals.

Although there was considerable progress in the design of all the products and in technical product innovation, the working lives of people employed in the new factories were not fundamentally different from those of people working in the older industries except that the work was generally somewhat lighter, cleaner, and quieter; and morale was higher because of the excitement of being involved in something new and growing. In the distributive trades and in food and drink processing the traditional small local firms were still predominant; and although this period saw the development of retail chains such as Marks and Spencer and Woolworths, and many other large department stores in cities, there was nevertheless a growth of small shops such as tobacconists and confectioners, and others which specialized in the sale of the new domestic goods such as radios, gramophones, washing machines, vacuum cleaners. Consequently a great deal of retail business remained in the hands of small – often family – firms working in a very traditional way.

It is mainly since the end of the Second World War that working life in Britain has been transformed by technological developments. The impact has been particularly severe in this country for several historical reasons. Because of its imperial past, and its pioneering position as the first industrial nation, a very abnormal and distorted economy had developed. The traditional heavy industries and textiles of the north had been built up on a vast export trade, much of it to the Empire, which at its peak accounted for half the international trade of the world. As foreign competition in international markets inevitably increased, these oversize companies were weighed down by the conservatism born of success in an earlier age, combined with obsolescent plant and methods: they became industrial dinosaurs. During the thirties free trade faded and Imperial Preference prevailed, so that it was not necessary for British firms to modernize products to world market standards, nor to improve productivity to internationally competitive levels. One vitally important consequence of these historical factors is that much of British industry has not yet gone through the transition from 'producer orientation' (making goods in the hope that they will be bought as produced) to 'consumer orientation' (designing products to satisfy the needs and interests of consumers). The British imbalance problem has been further compounded by a continuing overemphasis on export trading, with inadequate attention to domestic markets; and by a disproportionate commitment of scarce technical resources (engineers, technologists, and scientists) to defence and to the nuclear power and aviation industries. Vitally important producers such as the machine tool and automobile industries have suffered severely as a result of inadequate technical investment.

In agriculture specialized cultivating and processing machinery, combined with equally important science-based developments in plant and animal breeding, has transformed the lives of farm workers and led to extensive rural depopulation and the destruction of much of traditional village life. Most of the manual labour on farms has been replaced by machine power, so that agriculture accounts in Britain for only about 3% of total employment – only half the proportion in other European countries, albeit producing more than 70% of the nation's food needs. With the sole exception of shepherds, modern farm workers are as much skilled operators of machines as are factory hands. However, because they are so few and thinly spread, they are required to undertake all of the wide range of tasks that have to be performed. In that sense they are probably the most skilled of all industrial workers; and despite the fact that their productivity (added value/per worker) is higher than that of any other category of industrial worker except those employed in oil and chemicals, they are by far the worst paid. The quality of their work and the satisfaction it gives is, however, proportionately higher than for many other industrial workers.

The rapid mechanization of agriculture has brought about a big increase in the size of farm units, in the same way that technical developments have greatly increased the size of many other industrial plants and firms. The influence of technology on the size of enterprises has been one of the main factors in changing the working lives of employees. It has been accompanied by another trend, to greater complexity. This is well illustrated in the food and drink processing industries. Between the wars much of food processing was carried out in domestic kitchens and in small local processing plants (flour mills, bakeries, and so on). Fifty years later most of these have been replaced by big central processing plants, producing a vast range of packaged products, which are then distributed through a complex network fed by fleets of delivery vehicles of ever increasing size. This effect of technology on size and complexity has been greatly reinforced in Britain by the absence of significant controls on company takeovers and mergers. Indeed, for most of the last fifty years industrial conglomeration has been very much encouraged by successive governments. Technology and politically-inspired economic policy have been mutually reinforcing.

Throughout almost the whole of industry mechanization, and latterly some automation, has eliminated a high proportion of traditional labouring jobs and a fair proportion of semi-skilled and skilled production work. Mechanical handling systems and, recently, automated warehouses, mainly affected labouring work; increasingly specialized high-speed automatic machines were the main contributors so far as production workers and plant operators are concerned. Regrettably, much of the remaining production work becomes 'machine minding' rather than 'tool handling' – a dehumanizing process in that workers are increasingly subordinated to the plant and the speed with which it works. Motor-car factories have provided some of the most extreme examples, but the general effect is widespread.

Coincident with the reduction and change in the nature of shop-floor work there has developed a great expansion in white-collar service jobs, both inside and outside the main production companies, related to the production and distribution of goods. By 1971 the proportion of the total national work-force engaged in 'service occupations' had increased to nearly 50%, and more than half of these (27% of the total work-force) are engaged in one way or another in providing services to manufacturing and distributing companies, leaving only 23% of working people engaged in providing direct services to the general population; this percentage has been steadily decreasing despite the growth of government-funded services such as education, health, police, and social services. Compared with 27% of the total work-force engaged in services to industry and commerce, in 1971 only 19% were directly employed in the production of goods. Thus the traditional labouring, semi-skilled, and skilled industrial jobs are increasingly being replaced by various kinds of technicians, engineers, technologists, computer operators and analysts, as well as an increasing number of administrators and managers who are required to monitor, control, and plan ever bigger and more complex industrial operations. Some of the newer industries have become so technically demanding and capital intensive that a single country like Britain can no longer handle a major new project alone. For example, the development and production of the Airbus 320, planned to replace aircraft such as British Airways' Trident, requires a consortium of several European aviation manufacturing industries.

In the fifties and sixties the impact of technology on working life was felt mainly at the sharp end of production and distribution. In the seventies developments in electronics, computers, and other data processing equipment began to change the face of many offices, and the lives of those working in them. With increasing size and complexity the numbers employed in 'service' roles would have been even greater than is now the case, had it not been for the proliferation of such things as telex machines, photocopiers, and computers. There can be little doubt that a decline in the quality of a large part of office work is taking place as traditional jobs are increasingly replaced by the more fragmented work undertaken by a modern office machine minder. The general public experience some of the consequences of the dehumanization and demoralization of office work. For example, centralized and computerized accounting tends to obviate the need for a personal relationship between bank staff and customers: as staff morale deteriorates errors increase, and customers' confidence turns to irritation.

A similar change in the nature and satisfaction of work can be observed in supermarkets. No significant human relationship with customers is available to their employees, even for the cash register operators, whose work is so tediously repetitive that frequent breaks are necessary. (The supermarket operation is not yet, however, to any major extent a product of technological development. It depends primarily for its economic viability on the employment of unpaid customers doing much of the work traditionally carried out by paid shop assistants.)

The working lives of British coal miners have probably changed as a result of technological developments more than in most other forms of employment, as mechanical diggers and conveyors took over from pick and shovel and pony-drawn rail wagons. Even the traditional wooden pit-props have been replaced by sophisticated hydraulic roof-supports. The comparatively few remaining face-workers have become skilled machine-operators.

On the railways mechanical handling has done away with a great deal of traditional work, while the skills required by the driver and fireman have become redundant as steam trains were replaced by diesels, which in turn are giving way to electrical propulsion.

Although there have been very significant changes in building design and materials, attempts to industrialize building construction have largely failed. As a result, although some of the labouring work has been replaced by mechanical handling equipment on building sites, most of the traditional trades remain basically unchanged.

Thus over the period of the last three decades working life in almost all types of job, in nearly all the main places of employment, has radically changed, to a considerable extent as a consequence of technological innovation both in products and in the processes of production and distribution. A large proportion of unskilled labouring jobs have gone forever: many traditional skilled and semi-skilled jobs have either disappeared or quite new and different skills – often in service or managerial roles – have taken their place. The nature and the quality of work has changed; sometimes for the better, sometimes it has deteriorated. To those working in traditional industries which are in decline, or are competitively unsuccessful, it inevitably appears that technological innovation destroys employment. The picture in new or growing industries looks entirely different: here innovation is the main provider of jobs. The question of the overall effect on total employment opportunities cannot be answered in general terms, because it depends upon the overall dynamism of particular national or regional economies. Despite the currently high level of unemployment in Britain, in excess of three million, there are more people in employment in 1986 than there were at the beginning of the post-Second World War period considered here. So it cannot be said that technological change has reduced overall employment in Britain, even though it has been a main cause of change in the nature, quality, and location of most forms of work.

Domestic life

Although within living memory working life has been transformed to a considerable extent for most workers as a result of technological innovations, the home lives of the population have undergone an even greater metamorphosis. The extent to which this is a result of technological factors is hard to overestimate.

The average net per capita wealth creation within Britain almost trebled in real terms between 1945 and 1983. For the better-off 80% of the population that meant a degree of considerable affluence in which high standards of housing, diet, home heating, health care, and education to secondary standards were enhanced by equally high standards of entertainment and of cultural and leisure opportunities. Even for the poorest 20%, many of whom still suffer considerable relative poverty, living conditions are vastly superior, even in a period of recession, to those that existed between the wars and before 1914. These material improvements are the result of many interacting factors, but there is no doubt that the two basic elements have been technological innovation and a unique historic period of low-cost fossil energy.

These two elements influenced change in domestic life by means of two basic mechanisms. Firstly, the increase in manpower productivity at a compound rate of about 3.5% in manufacturing industry (and even more in agriculture) over the whole post-war period released many people for alternative forms of employment in a widening range of industries and services. The resulting higher family incomes expanded the capacity of most people to purchase the goods and services produced by the new employment. For example, during post-war years much of the domestic kitchen work in the preparation and processing of foods was industrialized. Indeed, a significant part of economic growth was the result of a transfer of activities from the home into the formal money economy. Higher incomes, resulting from greater wealth creation and increased manpower productivity, enabled people to purchase goods and services rather than provide them for themselves at home.

The second way in which technological innovation effected change in domestic life resulted from the development of a wide range of household products specifically designed to help provide a very high standard of material comfort and convenience in the absence of any significant amount of paid domestic service. A low-cost supply of electrically-powered appliances such as automatic washing machines, spin driers, and tumble driers, and synthetic detergents, have transformed the laundering of clothes and household linen and almost completely eliminated the labour traditionally involved. Synthetic fibres and new treatments of textiles have reduced ironing to a minimum and, with enhanced wear properties, a considerable amount of repair work on clothing has disappeared. Automatic space and water-heating systems have virtually eliminated the daily chore of cleaning fire grates and stoking fires, as well as providing a much higher standard of home heating. A great deal of the work of cleaning in the home has been lightened as a result of new high quality cleaning materials and machines such as vacuum cleaners and dishwashers. The use of new synthetic materials in flooring, carpeting, and surfacing has contributed to easy maintenance.

The combined effect of all these development has been a very considerable increase in labour productivity in the home. This has provided greater opportunities, particularly for married women, for increased leisure, more time for enjoying family life, and freedom to engage in paid employment.

Family life and leisure

Until 1945 the standard working week was 48 hours – plus often as much as 10 hours' overtime – over 5 1/2 days; and in addition to Bank Holidays there was only one week's paid annual vacation. With little money to spare after necessities had been paid for, there was neither time nor cash available for anything but a very restricted family life and leisure activity. The higher incomes and shorter working hours, in the home as well as at places of employment, that developed during the last three decades have transformed family life and leisure for almost the whole population. The standard five-day working week is now less than 40 hours in most occupations. Four weeks' annual paid leave is fairly common and few jobs offer less than three weeks. Holidays abroad, and most of the sports and recreational activities that were exclusive to the wealthier 10% of the population, are now almost universally enjoyed.

With almost every household possessing a radio and television, everyone has access to a range of entertainment, sport, and culture that was previously not available to even the wealthiest members of society. With almost half the households of Britain owning a car, and with a comprehensive public transport system, travel and the enjoyment of all the rich treasures of the nation – natural and man-made – have become almost universally available.

Leisure clothing, almost unavailable for most people before the Second World War, is now commonplace, adding pleasure to those occasions on which it is used.

Last, but by no means least, the extension of education and the National Health Service has been dependent upon the increased productivity and wealth creation of agriculture, industry, and the private service sector of the economy.

The price paid

The Americans have a saying, 'There's no such thing as a free lunch'. It would be hard indeed to find any change in human life in which there was no single disadvantage: the best we can hope for in making changes is to gain more advantages than disadvantages. A formidable catalogue of benefits has been listed in the foregoing paragraphs, and little has been said of the price that has been paid for that undoubted progress. In examining this price, we must initially do no more than identify particular significant disadvantages that have been incurred. Later chapters of this book will consider alternative directions of development which may provide the same or similar benefits at less cost. Only then will it be possible to answer the important question of whether too high a price has been paid through the means (technologies and economic policies) that have brought us to where we are at the present time. The gains that have been made so far are so substantial for a great proportion of the population that any departure from this well-tried approach will need to be supported by a conviction that the cost is indeed such that alternatives must be sought, and that there are practicable alternatives which offer the prospect of a substantially reduced cost.

It would be insensitive to begin this examination of the price that has been, and is being, paid without first mentioning that poor section of the community which has gained least in relative terms. At the end of the Second World War there existed a wide gap in incomes between the extremes of the social spectrum; despite steadily increasing levels of taxation only a slight narrowing has occurred in Britain between the top earners and the average, and there has been no change at all in the gap between the lowest incomes and the average. Policies for income redistribution have been an almost total failure. One important reason for that failure is that capital-intensive methods of wealth creation have been the basis of post-war economic growth. This has had a dual effect. To pay for increasing capital investment more of the new wealth creation was channelled to the owners of capital, providing the richer section of the population with a cushion against redistribution. And employees in those industries which achieved increased productivity by adopting new technologies were able to gain higher wage increases than people in other jobs, becoming the pace-setters in wage escalation. Many labour-intensive jobs disappeared as a result – for example, a great deal of house painting and maintenance became part of the unpaid economy ('do it yourself', or DIY) – and those that remained became the low pay sector. The major role that capital-intensive production technology has played in preventing effective income redistribution in favour of the poor has been amplified since the development of high rates of monetary inflation in the 1970s.

Another effect of capital-intensive methods which must now be counted as a recent cost is the reduction in demand for labour, particularly unskilled and semi-skilled workers. While economic growth rates were similar to rates of productivity increases in the fifties and sixties, displaced labour could be re-employed, particularly in expanding service industries. With further increases in labour productivity and very much reduced rates of economic growth in the seventies and eighties unemployment has increased dramatically.

The two effects of capital-intensive methods – income differentials and unemployment in times of low rates of economic growth – are costs that will need to be carefully examined in the chapters that follow.

The poor section of the community is also disproportionately affected by another price demanded by the existing system. The past three decades have seen the first abandoning in the history of mankind of a 'conserver' ethic. Not long ago even items of clothing were passed from one generation to another; and many buildings from the seventeenth and eighteenth centuries remain, while some modern buildings are being pulled down or are becoming uninhabitable after little more than a decade. So called 'durable' goods – cars, cookers, washing machines, and so on – are thrown away after as little as ten years. This is the age of 'consumerism', an age of almost unimaginable waste. For example, over an average lifetime each British citizen now uses about 400 tonnes of glass, most of which ends up as waste within the same period of a lifetime. In every system there is some unavoidable waste, but it is obvious that much of present waste is unnecessary and is avoidable, given the operation of a 'conserver' rather than a 'consuming' system. In every purchase a certain amount of 'waste' is paid for. When some of that 'waste' is unnecessary it is a cost burden on the purchaser which bears most heavily on the poorest in society. Although a reduction in waste would have no significant effect in reducing relative poverty, the scale of waste in Britain is now so great that an attainable reduction could probably do more than any other single thing to increase the living standards of poor people. For example, with well designed buildings and heating systems energy consumption can be reduced to about a quarter of normal British levels. In poorly insulated buildings fitted with traditional heating systems many poor families are being impoverished by the high cost of heating. This situation will become unbearable if, as is frequently predicted, the real cost of energy doubles during the next few decades.

There are two other prices associated with the unnecessary waste of the present system – consumption of non-renewable natural resources and the harm done to the environment. Britain has good cause to be particularly concerned about both. With the exception of coal, oil, and natural gas, we possess very few economic sources of industrial minerals; there is insufficient agricultural land to provide the present population with a traditional diet; and we have less forestry than most other industrial countries. We are consequently heavily dependent upon imported raw materials and food, for which we have to compete with many other countries, some of which are economically much stronger than we are. This dependence is becoming an increasing burden in a world with a rapidly increasing population and limited resources, of which many important ones will be reaching the peak of their production within the next few decades. Unnecessary waste of scarce resources is a recipe for bankruptcy.

A great deal of environmental damage has been caused by the comparatively uncontrolled disposal of industrial, agricultural, and domestic waste. Rivers and beaches have been heavily polluted, and untold damage has been done to buildings and public health as a result of air pollution and noise. There is also increasing concern that irreparable damage is being done to our most precious resource – the fertile soil – by industrial farming methods which maximize short-term yields with little regard for long-term soil fertility. Nature is capable of safely absorbing a great quantity and a wide variety of pollutants, providing they are widely and thinly spread. Environmental damage is directly related to the concentration of emissions. Our modern society developed with highly concentrated areas of industrial activity and a population largely resident in major cities and metropolitan areas, the worst possible arrangement from the point of view of pollution. Ironically, now that the necessity of environmental control is accepted, the cost is at a maximum under present conditions.

In seeking ways of providing a high material living standard for the whole nation, we need to examine methods that are a great deal more efficient and less wasteful in the consumption of scarce resources. We also need to look for a more dispersed pattern of production and of habitat.

The consideration of concentration of activity and population leads to the next category of price that is being paid for our improved material way of life. Possibly the most dominant feature of economic development in Britain during recent decades has been the growth in size of most kinds of enterprise and in the scale and degree of centralization of production plant. For some industries, including coal, steel, oil, and chemicals, there are sound reasons for such development, and it is doubtful whether the benefits that have been gained in such industries could have taken place otherwise. It is by no means so evident that benefits have accrued from nearly three-quarters of the nation's bread being supplied by a handful of firms. If trends in the real prices of their products are used as a guide there would appear to be no public benefit from these concentrations compared with the products of the hundreds of small bakers that once existed. When we compare a large part of British manufacturing industry with overseas equivalents we see a much higher degree of concentration in Britain over a wide range of industries, and a comparatively small sector of small and medium-sized firms. Bearing in mind that many sectors of British industry have experienced difficulties in international competition at home and abroad, a question mark must be put alongside the generalized doctrine of 'bigger is better' for all types of activity.

The matter of firm or plant size has important human implications. The fragmentation and complexity that exists in many very big organizations, and the absence of opportunity to participate in decisions, is felt by many people to be dehumanizing. Although some people feel comfortable and secure in big organizations, most do not like to be a very small cog in a very big machine. As a consequence there are many more unresolved labour disputes in big firms than in small ones, and it is very much more difficult to maintain a high level of morale among employees.

Some of the main disadvantages resulting from technological change on the lives of working people have already been referred to in the section on 'Working Life'. It is also becoming evident that in an endeavour to remain internationally competitive in the marketing of materials and goods, and in some service functions, increasing work rates – people keeping pace with machines operating at higher speeds – is placing more and more stress on workers in all categories. The combination of unrelieved stress and unsatisfying tasks damages health and leads to increased absenteeism and alienation; so much so that the better-educated younger generation are questioning the worth of such forms of employment. The absurdity of a reducing number of people being required to increase work rates and suffer greater stress while an increasing number of their friends are unable to obtain any employment at all naturally reinforces the hostility of both employed and unemployed. For many people the price of keeping the existing system working is already too high.

The human price that is being paid extends well beyond the work-place. Family life is frequently damaged as a result of difficulties in working life, in some cases leading to marriage breakdown. An even greater social problem derives from the pressure to 'keep up with the Joneses', an inevitable consequence of a system based on the idea of maximizing consumption of bought goods and services. A great deal of nervous, emotional, mental, and physical suffering, often referred to as diseases of affluence, can be traced to this social pressure, and is present at all income levels. In a system that fosters acquisitiveness, to be excluded from paid employment is a very severe form of social deprivation. Although the material condition of the unemployed of the eighties is undoubtedly far less deprived than it was in the thirties, the psychological and emotional burden is now far greater.

Had we been concerned with an earlier period of industrial, technological, and economic development, up to, say, 1940, we should have identified only minor disadvantages to set against the benefits that had been gained. To a very large extent the price that has had to be paid for our present relatively high material living standard is one that has grown with the process of development since the end of the Second World War. The disadvantages are no longer minor: they have grown to such an extent that further development along the same lines poses a number of potential threats. We could become an increasingly divided nation between the 'haves' and the 'have nots' – the increasingly affluent and those who are increasingly deprived of material resources and are also without satisfying employment or a meaningful life. Persistence in 'consumerism' could lead to a further increase in environmental damage, and to starvation of essential raw materials and loss of food production, as scarce resources are priced out of reach. And as dissatisfaction with the absurdities of indiscriminate development spreads, demoralization and diseases of affluence could reach epidemic proportions. Because of continuing heavy dependence upon international trade – we have the highest proportion of imports and exports of all the major OECD countries – Britain is especially vulnerable and could rapidly find great difficulty in maintaining present living standards.

How changes have come about

It is obviously impossible in a few paragraphs to provide a comprehensive explanation of the various factors which have combined to change British economic life in the post-war years. Essentially, what has happened in Britain, as in other advanced capitalist industrialist countries, is that increasingly productive methods, product innovation, and a very cheap supply of fossil energy have been used to develop private and public sector markets for consumer goods and services. (Compared with socialist economies a free rein has been given in applying ingenuity in a competitive system to satisfying the needs of individuals.) However, there have been substantial differences in the way in which various capitalist countries have used the freedom of the market by applying different constraints through fiscal means, legislation, and regulatory controls.

For example, the rules applying to British agriculture gave no protection to small farms. Consequently a very advanced system of chemically based, highly mechanized agriculture evolved in Britain, supported by both government and private industry. In the outcome, British agriculture is markedly different from that of some other European countries, notably France. Although it is on a much smaller scale, it is more akin to American agriculture. The electricity industry in Britain sought to minimize its unit price by maximizing its generating efficiency in a nation-wide grid network. Small local power stations were replaced by centrally located plant of ever increasing size and efficiency, and no encouragement was given to industry to generate its own power. In many other countries a different approach was used. Big industrial plants generated their own power and used the hot-stack gases as a source of low-grade process heat; and individual cities built power stations and associated waste heat distribution systems for the space heating of buildings. In contrast to Britain these countries aimed to obtain the maximum benefit from the energy input of the fuel burned. A combined heat and power plant in Sweden would typically waste only 25% of its fuel energy: the British integrated system wastes nearly 75%. This and many other wasteful applications of energy in the British system are a cause of a much higher consumption of primary energy per unit of GDP that in most other European countries.

In the United States there have been for many years strict laws regulating competition. One effect has been that a very dynamic and prosperous small firms sector has been maintained, whereas the lack of such regulations in Britain allowed many such firms to be put out of business through the actions of bigger competitors. Local bakeries, for example, and food and drink processing companies, were absorbed and closed down, to be replaced by big, centrally located subsidiaries of large companies. The process of takeover and closure continued until there are now only two major companies producing almost three-quarters of Britain's bread, and only a handful of big firms cover a large part of the food and drink industries.

Fair trading regulation is another factor that has tended to favour small and medium-sized companies in some other countries. In the absence of such effective control in Britain most big firms are able to purchase goods and materials at very high discounts, compared with the prices paid by smaller firms. In many instances the total value of discounts is greater than the company's gross profit: in effect, inefficient big companies are subsidized by smaller firms who must be charged higher prices by their suppliers to keep themselves in business.

The concentration of economic power into a decreasing number of big companies results in the concentration of the technical resources of highly skilled engineers, technologists, scientists, and technicians. The imagination, inventiveness, and energies of these key people is directed to the development of products and processes which particularly suit the interests of a very big industrial or commercial firm. A result is illustrated by a new pharmaceutical factory in Britain: nearly all its equipment was obtained from overseas companies employing twenty people or less; hardly any was available from British sources.

The aggregate effect of the many factors which have affected the development of British industry differently from that of overseas competitors is shown by the way in which imports of foreign manufactured (including semi-manufactured) goods have steadily displaced British supplies to the home market.

It was once said that the interests of the United States were the same as the interests of General Motors. That may or may not have been true. It is certainly not true that what is good for a particular firm is necessarily in the public interest. Consequently, some degree of regulation is necessary if the public interest is to be served. It would appear that in Britain an unnecessarily high price has been paid by too much of the wrong kind of regulation being applied, and too little of the right kind. As a result, technology and the people who produce and use it have been misapplied.

Perceptions and expectations

In Britain, as in most other advanced economies, a small proportion of people are already almost completely disenchanted with modern technological society, so much so that they withdraw as far as possible from what is sometimes described as the rat race. But despite the many disadvantages, most of which have had a considerable public airing, there is little sign that the perceptions and great expectations of the vast majority of people have undergone any appreciable change. Almost forty years of exposure to promises of improvements in material standards, matched for the majority of people by realization of those promises, has produced an understandably firm belief in the system of economism/consumerism that has 'delivered the goods'. Indeed, if support for 'green' political parties is any guide to a national mood for a radical change of direction in development, it would appear that, despite our greater vulnerability, there is less willingness here to adopt a new approach than in countries such as West Germany or France. Even the USA, the pioneer of the present consumerist system, seems to be more actively open to change than Britain.

Although the seventies and the first years of the eighties saw virtually no real growth in the British economy, expectations for the future remain as high as ever among most people. The call for zero growth falls on deaf ears. However, it need not be assumed that, given a choice, growth must inevitably consist of an ever-increasing consumption of material goods, energy, and food by all members of the population. Four-fifths of average household expenditure is absorbed in providing essentials – food and drink, housing, heat and light, clothing and footwear, and transport. These basic essentials absorb an even higher proportion in families with below average incomes, and in the poorest households almost all expenditure is on these items. At our present level of development only a small minority of the population would gain from a higher consumption of food, drink, and energy. While there is still a need for more housing, it is increasingly being met by upgrading existing property rather than by more material-consuming new construction. The only item that is likely to increase substantially in this list is transport, as the half of the population who do not own cars become motorists.

There is a fundamental difference in perception between ordinary people and economists about how increasing expectations are to be satisfied. Most people, once they have obtained a sufficient quantity of food to keep them healthy and satisfy their hunger, do not eat more and more as their income increases. Instead they change their eating habits, choosing more expensive and better-quality foods or eating in restaurants rather than at home. This 'quantitative saturation' effect applies to all forms of consumption: at a certain stage of economic development the emphasis shifts from quantitative increases in consumption of both goods and services to improvements in quality. In Britain probably 75-80% of the population are now more interested in improving the quality of their lives than in merely having more of everything. It can confidently be predicted that national growth in four-fifths of the economy will consist mainly of qualitative improvement rather than quantitative increases in consumption. Since it is consumption which is most material and energy intensive, the use of resources is unlikely to increase significantly, and with increasing need and pressure to become more efficient in their use, overall consumption of materials and energy would most likely fall. If improvement in the quality of life can be obtained with less consumption of materials and energy, ordinary people are not concerned; and if it can be obtained at less financial cost they are delighted. If I can provide myself with a comfortably warm house using different heating systems, and as a result my heating bills decrease, I am very happy.

Those whose perceptions of economic development are conditioned by conventional economics see matters in a different light. Changes that result in a lowering of consumption, including less expenditure of money, are considered to be, without distinction, lowering the standard of living. 'Economically' the standard of living can only increase if there is an increase in overall consumption, expressed in monetary terms. In the world of the economist quantity rules and quality is only significant if it is represented by a price increase.

The gigantic amount of waste in the existing system implies the possibility of providing for our needs and satisfactions in much less wasteful ways. A change of direction in technological and economic development towards more efficient systems could increase living standards without increasing consumption. Were this to happen the perception gap would widen between ordinary people and those observing the rules of traditional economics.

The Saint-Geours Report to the EEC on An Energy Efficient Society recognized that such a different society cannot simply be developed from a planning office. It requires a high degree of individual initiative and responsibility if people are to achieve their aim of a better life.

Where are we going?

Suppose that a change of direction in development does not take place; and the rigidities of the over-concentrated, over-centralized British system and its institutions provide plenty of reasons for thinking that a radical change will be extremely difficult. What then is likely to happen? Where will we be going?

Nothing very dramatic is likely to happen in Britain so long as North Sea oil and gas hold out, providing us with protection from the otherwise drastic consequences of our declining manufacturing industry. As oilfields in current use become exhausted, and the number of new fields to be discovered declines, output could drop sharply over a decade or so. Nobody can be sure when that will happen, but it is unlikely to be more than thirty years away. Britain will then be in a very sorry state if nothing is done now to change the direction from consumerism to conserverism.

With no change of direction and a continuing decline in industry – a continuing fall in the share of both the British market and worldwide overseas markets – the loss of oil and gas revenues will have a grave effect on the ability to maintain free public services, which by then will inevitably be at a substantially lower level. It is very likely that virtually all the real gains of the past thirty years in national living standards will be lost, and the situation of the poorest people in the country, who have not benefited from income redistribution since the Second World War, could be even worse than it was in 1945.

Protagonists of the view that there is no workable alternative to the present consumerist system believe that if British industry becomes more efficient the future prospect is good. They seem not to have noticed that even though the least efficient parts of British industry disappeared in the late seventies and early eighties, and those remaining have improved their labour productivity, the flood of imports has continued to rise, so that for the first time since the first industrial revolution the value of manufactured imports exceeds that of exports. Neither do they seem to have noticed that other more efficient European countries, such as West Germany and France, also find their situation worsening as they face increasing competition in the hitherto soft world markets that have accounted to a considerable extent for their prosperity from the newly industrialized countries. As more and more countries become industrialized and enter into world trade the proportion of trade enjoyed by the countries that developed early will inevitably decrease. Although the total volume of world trade may grow, it is highly unlikely that the rate of increase will be sufficient to balance the loss of market share that Britain will suffer. Because of the still exceptionally heavy dependence of the British economy on overseas trade, Britain is extremely vulnerable in the face of the new competition: moreover, the nearly saturated European markets account for a large proportion of British exports.

Dr E.F. Schumacher described those who put their faith in a continuation of the consumerist system as 'supporters on the onward stampede'. There is no sign of optimism among them that Britain is likely to recapture the lost ground in traditional industries like textiles, iron and steel, shipbuilding, machine tools, and other machinery. Reliance on the newest industries, such as microelectronics and biotechnology, is their hope for the future, despite the fact that even the most optimistic expectations for these industries pale into insignificance when compared with the erstwhile output of traditional industries. Even the capture of an exceptionally high share of these new product markets could not begin to offset continuing losses in traditional business.

An important long-term consequence of no change in the economic approach in Britain is the steady rise in prices of raw materials and energy. The wasteful consumerist approach has only been possible because there was access to very low-cost fossil energy and imported raw materials. As worldwide demand increases and production of some important industrial minerals (as well as oil and gas) passes its peak, price increases will ultimately bring the consumerist era to an end. Those who have prepared themselves well in advance for that development will make the most comfortable transition to an alternative system.

The dire prospect for Britain, if it fails to make an early adjustment, is unthinkable for future generations. There is an urgent need to look for a more appropriate development approach to meet the needs and expectations of ordinary people.

Where should we be going?

Since the availability of cheap energy has been a key factor in achieving a high quality of life in the developed countries, similar standards for the world population cannot be attained unless a new technological system can be produced which is much less demanding in energy input. About three-quarters of all industrial energy is consumed in the production of materials such as metals, plastics, cement, glass, and so the new technological system must also greatly reduce the consumption of materials.

In order to grasp the magnitude of change required, energy consumption can provide a useful guide. If escalating energy prices, resulting from demand exceeding supply from cheap sources, are to be avoided, estimates indicate that during the next half-century average world-wide per capita consumption should not exceed about 1.5 tonnes of coal equivalent (tce) per annum. Since tropical countries do not have a significant space heating problem compared to those with colder climates, their consumption should be lower, so that a rate of energy use in Britain of 2 tonnes of coal equivalent per person per annum would be acceptable: the current rate of consumption is be tween 5 and 6 tonnes of coal equivalent. The target is therefore to maintain the present quality of life, with improvement for the poor, with an economic and technological system that consumes only about one-third of the energy used at present.

It is not too difficult to see how such an enormous reduction could nearly be achieved in terms of delivered energy. Figure 1 shows that 60% of all delivered energy is in the form of heat, two-thirds at a low temperature mostly used for heating buildings. By a combination of

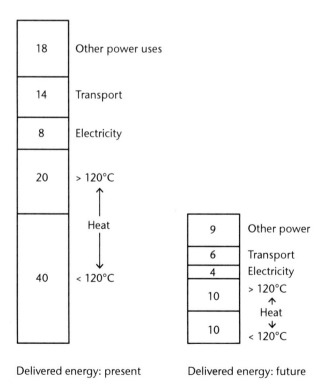

Delivered energy: present   Delivered energy: future

Figure 1 Delivered energy in the present and future

greatly increased insulation, improved ventilation, the use of heat pumps instead of direct-fired boilers, the use of waste heat in combined heat and power (CHP) systems, and the judicious application of passive solar heating devices, the heat required for buildings could be reduced by a factor of 4. Nearly all the higher temperature heat (more than 120°C), half the electricity, and all other power uses are taken up by industry and commerce – nearly three-quarters being used in the energy-intensive materials producing sector. By increasing repair, reconditioning, re-use, and recycling of durable goods (the 4Rs), energy consumption would be halved as average life was doubled. For example, average car life would be increased to 20 years: in Sweden they have already reached 17 years.

Britain's electricity consumption per capita is almost as high as in countries with double the GDP. As well as halving industrial consumption, substantial reductions in consumption of electricity could be made possible by increasing efficiency of use in most applications. Overall electricity consumption could be halved if full advantage were taken of all available opportunities. Transport delivered energy can be reduced by almost a factor of three. Consumption of energy in transport in Britain is exceptionally high because of the very high proportion (90%) of freight carried by road, where the energy required to carry a tonne for 1 mile is about three times as high as for rail or water transport. An equally high proportion of passenger transport is also by road, most of it being in cars with fuel consumption in the region of 30 mpg. Cars of the next century will triple their efficiency: claims of 90 mpg are already made for a Renault experimental car.

If delivered energy can in principle be reduced to about 40% of the present level, the reduction is even greater when this is translated into primary energy (coal, oil, gas, hydro, nuclear, wind, solar biomass). (Electricity makes a substantial contribution to this further decrease in consumption, because it takes about one-third of total primary energy to provide the 8% of delivered electricity.) Thus, the overall primary energy requirement should be little more than one-third of present consumption for the same quality of life – that is 2 tce/per capita/per annum instead of S-6 tce.

No sceptical person could accept such a thumbnail sketch unless it could be elaborated in much greater detail. Such an elaboration has been presented by Gerald Leach in his book, *A Low Energy Strategy for the UK*. It is a truly practicable possibility. Additional substantiation is to be found in the World Energy Conference study on energy conservation and efficiency. In 1979 they forecast a 47% reduction of energy consumed worldwide per unit of GOP over the following three decades. Because of Britain's performance at that time – more energy consumed (and wasted) per unit of GDP than in most other industrialized nations – an even greater saving is feasible in Britain than the worldwide average. Although a low-energy, low material-consuming system is an essential part of a new direction of development, and the substitution of the 4Rs (repair, reconditioning, re-use, and recycling) for a proportion of manufacture is a key element in it, there are other requirements to be fulfilled if the price of maintaining a high quality of life is to be brought down to an acceptable level.

There are clear indications that most people would prefer to live in small or medium-sized towns rather than metropolitan areas and big cities; work in small or medium-sized firms is more satisfying for most people; and almost everyone wants paid employment sufficient to be independent of social provision. Ultimately, full employment can only be restored by sharing all the paid-for work that is available, and it must be work that is designed to suit all the various types and talents that make up the population. As yet there is no sign that this has been widely accepted. However, the need for dispersal of population and the move to smaller units is recognized and it is beginning to occur. National census figures confirm the former, while remarks made by Sir Adrian Cadbury, Chairman of the Cadbury Schweppes Group, illustrate the latter. In a *Guardian* article in 1981 he wrote:

> To achieve industrial survival and international competitiveness [in Britain] means reversing the trend of the last twenty years towards large, centralised organisations. We will want to break these organisations down into their separate business units and to give these units freedom ... because big is expensive and inflexible.

Britain has a lot of leeway to make up, as will be seen by comparison of mini steel mills, some with an annual capacity of less than 100,000 tonnes:

| Japan 64 | Italy 62 | USA 50 | Spain 25 | UK 7 |

If applied discriminatingly, the new microelectronic technologies can powerfully assist the development of this more humanly satisfying, less environmentally damaging, less capital-intensive form of economic and social system. Such a radical process of change will inevitably meet with strong resistance from the many vested interests which especially benefit from a concentrated, centralized, consumerist society. As the Saint-Geours Report said of an energy efficient society, 'it requires a high degree of individual initiative and responsibility' if ordinary people, everywhere, are to achieve their aim of a better life.

CHAPTER 2

Manpower, money, and machines

The notion that has dominated the approach to economic development in Britain since the Second World War is that progress can only be made if it is based on a high rate of increase in industrial productivity. The key to that increase is generally believed to be an increase in the scale of industrial operations, with fewer companies satisfying greater demand, combined with increases in capital intensity. The underlying beliefs are

1. that bigger is always better;
2. it is more effective and efficient to employ money and machines rather than people.

These are beliefs that had their origins in heavy industry. They then spread over the whole industrial spectrum and have subsequently taken over most economic activities – distribution and retailing, schools, hospitals and banking. They are beliefs which are deeply embedded, so that it is difficult for many people to envisage any credible economic development alternative that is inconsistent with them.

The purpose of this chapter is to challenge these basic beliefs about how to achieve the most effective mix of men, money, and machines. It is not an attempt to demonstrate the opposite of these two convictions (i.e. that small is always better, and it is always better to employ people rather than machines and money). Rather it suggests that the commonly held beliefs in favour of size and capital intensity are exaggerated and over-simplified. Undoubtedly there are 'economies of scale'; but there are also 'diseconomies of scale'. The economies tend to diminish as size increases, whereas the diseconomies become greater. Consequently for each individual activity size needs to be optimized rather than maximized.

There are certain industries, notably process industries, in which to be effective and efficient a high level of capital investment is necessary and comparatively few people need to be employed. But in most other economic activities it is practically possible to achieve a high level of overall resource productivity with a variety of mixes of men, money, and machines. In the case of capital/labour intensity, maximization of capital can be disastrous if indiscriminately applied; and for any particular level of economic development there is for most activities a spectrum of mixes of men, money, and machines that are equally effective.

How the industrial system has evolved in Britain

Before the development of the steam engine at the beginning of the industrial revolution there was very little investment of money in the production of goods. Apart from windmills and waterwheels, used primarily for milling flour and manufacturing textiles, production depended upon hand tools. When a fleece was converted to yarn, then to cloth, and finally to a garment, the only costs incurred, apart from the price paid for the fleece, were the wages earned by the people who operated the spinning wheel and hand loom and those who sewed the garment. The price paid by the buyer of the garment amounted to no more than the cost of the raw material (the fleece) plus the wages of the several people who, with their special hand skills, 'added value' by converting useless material into a useful product. In that primitive system there was no need for an entrepreneur to stand between skilled workers and the buyer, because there was no need for investment other than the workers' skills applied through their own tools of the trade. Needless to say, some master craftsmen employed assistants. In addition to the wages they earned directly from their own work, they made a profit on the work done by assistants whose pay was less than their own. With the exception of that profit, the price paid for a manufactured article was directly related to the number of hours spent in its production and the rate per hour paid to skilled workers.

The introduction of powered machines broke that simple pattern. By investing in a factory full of machines, an entrepreneur was able greatly to reduce the number of man-hours taken to do a job. This enabled him to sell the product at a lower price than was necessary to cover the bare costs of hand-made products, initially leaving a handsome profit on his investment. As competition with other factory-made goods developed, prices began to fall, and in order to try to maintain his profit level he often resorted to cutting the wages of his employees. Having started at mere subsistence levels of earnings, workers soon found themselves being exploited and impoverished in the new factory system.

When finished product prices were reduced the amount of added value was reduced. ('Added value' is the difference between the selling price and the cost of raw materials and other purchased items that are necessary in producing a finished product.) This was the beginning of competition between manpower costs and return on investment – the cost of finance – as employees and employers sought to maximize their share of the added value.

With the passage of time, and with machines capable of producing more goods in less time and with fewer employees, the production cost of most goods has been reduced. Whether added value increased or decreased depended on changes in the market price of the article being produced. Quite often competitive pressures on market price were so great that the added value on each item fell. When that happened, the competition between employees and employers to maintain their income was intensified. Investment in new and even more productive technology – or more intensive use of existing machines – was one escape route for management when confronted by determined wage demands.

In most modern business the claims on added value are no longer simply between workers and equity shareholders: many other parties are involved.

Added value – who gets what share?

All commercial and industrial activity is concerned with the creation of wealth, or 'adding value'. At each step in the process of producing and delivering a product or service to a consumer certain costs are incurred. All kinds of purchases must be made in a modern business – materials, components, packaging, energy, etc. – and suppliers of labour, property, services, finance, and so on, as well as shareholders, must be paid. In the long run, income from sales must cover all these charges if a business is to remain in being. The difference between sales income and the total cost of all these purchases represents the quantity of added value; and this quantity must meet all manpower and financial costs, including such taxes as may be levied on profits. The sharing of added value between those who supply their skills and labour (manpower costs) and those who provide finance can be identified from Figure 2.

Although there are considerable differences between companies and industries in the proportion of sales income appearing as added value it is rarely more than 50% and seldom less than 25%. An added value of about one-third of sales income is fairly common. The split of added value between manpower and finance costs also varies considerably. In a very labour-intensive business with little fixed or working capital, manpower costs may be little less than added value. On the other hand, in an industry such as oil or chemicals where the capital investment is very high and the workforce comparatively small (capital employed per employee may be several hundred thousand pounds), financial costs take a major share of the added value.

Although the concept of manpower costs is commonplace, it is not usual to consider the aggregation of rent and loan finance interest, taxation and the net yield after tax on the equity shareholders'

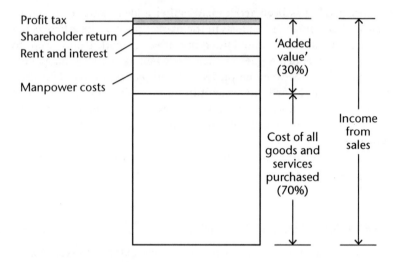

Figure 2 Share of added value between manpower and financial costs

investment, as a cost. Strictly speaking, only the rent and loan interest are in the form of bills that have to be paid. The rest is net profit, which has to be allocated for the three purposes of reinvestment in the business, equity shareholders' dividend, and taxation. The aggregation can, however, be said to represent a notional cost of the total finance invested to enable all the people engaged in the business to add value. In every business there are only two categories of input – people and money; and value is added by mixing manpower with money and with machines that cost money. Because some of the suppliers of finance – for rent and loan interest – enjoy a fixed return, pay negotiations are aimed at dividing only that part of added value which excludes rent and loan interest.

As businesses have become increasingly capital intensive, increasing amounts of money have had to be found out of added value to remunerate investors. But this has not meant that employees have over the years received a decreasing share of the added value created. Although it has been additional investment in more productive plant that has increased the amount of added value per employee, it is employees whose share of added value has increased and not the share taken by investors. Over many decades the real value of employees' earnings has increased as equity investors' profits have gradually decreased. This was particularly marked during the first half of the seventies when there was serious overmanning in many British companies: between 1972 and 1975 wages and salaries as a percentage of added value increased from 68% to over 80%. In Japan in the same period, by comparison, the rise was from 47% to 51%, thereby allowing the Japanese a much higher level of reinvestment. With reduction in overmanning the British wages/salaries proportion has declined somewhat, but it remains high in comparison with countries where industrial reinvestment has been greater. The way in which added value is shared in Britain is unlikely to change very much in future; there is no room for the share taken by wages and salaries to increase further in a high capital system, and it is very improbable, now that overmanning has been markedly reduced, that trade union negotiators will allow the share to decrease significantly. In so far as there may be a possibility of operating efficiently (high productivity of both manpower and capital) with relatively modest levels of capital intensity (capital/employee) it would appear that such a course would be particularly appropriate to the way in which added value is shared in Britain.

Ideas of productivity

The idea that productivity is measured by the amount of added value created by a single employee of an organization has its roots in the age of hand tools and primitive machine tools, whose speed of production was determined by the operator. Most modern industrialized systems of production and distribution have output rates largely determined by the technologies used, not by the effort exerted by employees. In so far as employees have an influence on output rates it is in the proportion of working time during which plant is in full production. Much of the criticism levelled at low productivity in Britain has been concerned with people, as though hand tools were still in vogue; in fact it has been caused by comparatively poor plant utilization and by overmanning (mainly in overhead manning levels rather than in direct plant operator manning levels) for both of which management must be held responsible.

In a modern industrial system it is obviously misleading to consider productivity solely in terms of output, or added value, per employee when the level of capital investment is often a dominant factor. To obtain a practical idea of productivity, the inputs of manpower and of investment need to be combined. Figure 3 provides a useful picture.

The line AB represents constant overall resource productivity of 100, with different mixtures of manpower cost and total investment cost at each point on the line.

The line CD (50) represents a higher level of overall resource

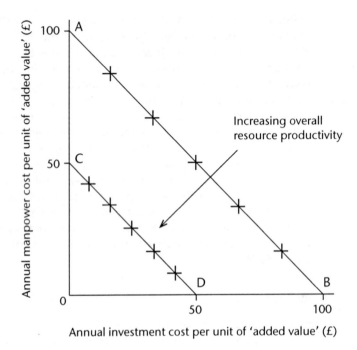

Figure 3 Resource productivity with various mixtures of investment and manpower

productivity than the line AB (100) because smaller quantities of manpower cost and investment cost are used to create a unit of added value. Overall resource productivity increases as the line approaches the zero.

Figure 4 can be used to illustrate the trend in the changing mixture of capital and labour cost as, for example, hand looms are replaced by power looms in the production of cloth. Point A represents hand loom operation when all added value covers labour costs only.

Point B: investment in power looms increases overall productivity. The investment cost is more than offset by reduced labour costs as output per hour is doubled.

Point C: no change in machines; but wages are reduced as a result of market competition.

Point D: change to a two-shift system reduces capital costs per unit of added value with no change in wages.

Point E: more productive high-tech looms reduce both capital and labour costs per unit of added value, even though wages and the return on capital are increased.

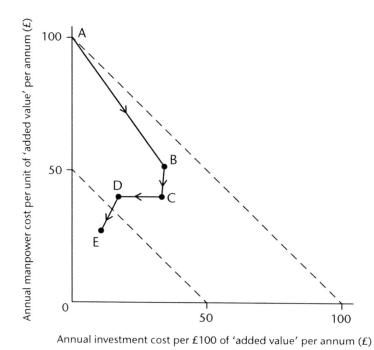

Figure 4 Changes in the relation of capital and labour costs in manufacturing

It will be seen that overall productivity increases as each point moves closer to zero.

In practice, however, the same level of overall productivity can be obtained in any particular industry with different mixes of manpower and money, using different technologies and different scales of operation.

In his James Clayton Lecture to the Institution of Mechanical Engineers entitled 'The Economic Ingredients of Industrial Success', Dr F.E. Jones published comparative performance data between British and Japanese companies. Using his data the following comparison can be made between the aggregation of UK manufacturing industry in 1976 and 416 Japanese manufacturing companies in 1974.

Per £1m of 'added value' per annum	UK	Japan
Number of employees	204	130
Cost per employee/annum	£3,500	£3,500
Total capital	£1.5m	£3.8m
Fixed assets	£0.92m	£1.49m

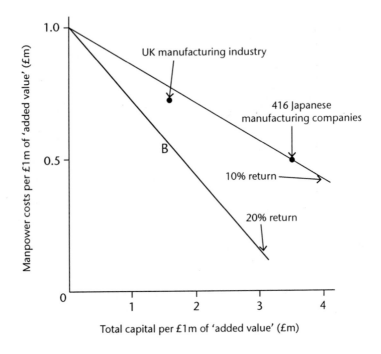

Figure 5 Comparing combined labour and capital productivity of UK and Japanese manufacturing

With the same level of average employee remuneration we see two markedly different mixes of manpower and capital being used in the two countries to produce £1m of added value. Figure 5, plotting manpower cost against the total capital employed, shows two lines representing 10% and 20% return on the total investment. Although the UK mix of men and money involves much less capital and many more employees to produce the £1m of added value, it nevertheless produces a slightly higher return on the total money invested. Since at the times concerned employee remuneration levels were the same in the two countries, it can be seen that the overall utilization of resources was slightly better in the UK than in Japan. Because the cost of money in the UK is usually considerably higher than it is in Japan, it would be counter-productive to attempt to operate the Japanese mix of manpower and money in Britain. This historic data was chosen for the purposes of illustration because, fortuitously, the average levels of wages/salaries were the same at that time in the two countries, and the capital intensities were very different. Since the mid-seventies many changes in manning, remuneration, and investment levels have occurred in both countries: no doubt an up-to-date comparison would look very different. However, the comparison does show that a high capital/manpower cost ratio is not necessarily more productive overall than a mix at a lower capital level. Indeed, the highest overall levels of productivity are obtained when both the capital investment and the manpower cost are minimized in the creation of added value. The optimum mix at any particular time will be different in different countries, depending upon levels of the unit costs of manpower and money and on the effectiveness with which it is possible to use both resources. For example, a country where money is cheap, labour rates are high, and plant can be kept in operation twenty-four hours a day, seven days a week, is likely to have an optimum higher in capital than a country where labour rates are equally high, but money is expensive and plants can only operate sixteen hours a day, five days a week.

The above international comparisons illustrate that the same total resource productivity is feasible over quite a wide range of different mixes of manpower and money. The same thing can be observed when inter-company comparisons are made in Britain within the same industry. The Morgan Car Company makes an interesting comparison with British Leyland and other high volume producers. Morgan, a specialist car producer, is very small indeed, having an output of about ten cars a week with a total workforce of about 150, using an absolute minimum of plant and machinery. The MG sports car was the closest BL model to the Morgan. Whereas MG went into liquidation, Morgan Cars continues to flourish because it is able to add value with comparatively low manpower and capital levels (i.e. its total resource productivity is good). It is interesting to note in passing that not only is Morgan Cars Ltd an efficient creator of wealth by traditional craft methods, but the Morgan 4 model has the lowest rate of depreciation of any car priced below £10,000 in the UK market. Consequently, this hand-made car is the cheapest car to own in Britain in this price range.

The efficiency of different-sized firms

The achievement of an alternative economic structure which is most humanly satisfying requires dispersal of the population into small and medium-sized towns, combined with employment in small or medium-sized organizations wherever possible. Of course there are a considerable number of people who prefer city life and many enjoy big company life, but a majority do not.

Historically all economic activity starts on a very small scale – Colonel Drake's oil well in Pennsylvania is recognized as the starting point for the modern oil industry; Ford and Morris, pioneers of the massive motor industry, began in what were no more than local garage workshops. Invariably, however, it was found that there were advantages to be gained as size increased. Nevertheless, all firms in a particular industry do not have to be roughly the same size in order to be competitively efficient: in most industries and commercial businesses a wide range of company sizes exists. It is much easier to indicate the activities that can only be done on a larger scale than it is to suggest businesses that can be done on a small scale.

It is mainly in the materials processing and production industries that it is seldom possible to operate on a small scale. North Sea oil, deep mining of coal, ethylene, cement, steel, and glass manufacture, are all activities which even at the smallest viable level are businesses of a considerable size. There are a few other product manufacturing companies, such as those that build large aircraft and supertankers, which necessarily have to be large scale. For the remainder there are companies and individual sites covering a very wide range of sizes. Even in an activity like vehicle design, development, and manufacture a few small firms, like Morgan Cars Ltd, survive; and in recent years one or two small aircraft manufacturers have been established. The composition in Britain in various economic sectors is indicated in the following table, which refers to firms employing 200 people or less.

	Small firms % of total (1963)		
	Employment	Value added	No. of firms
Manufacturing	34	43	7
Retail	28	21	46
Hotel/catering	11	10	16
Building/construction	10	11	10
Wholesale	6	1	7
Motor trade	3	5	5
Road transport	3	3	5
Mining/quarrying	–	1	–
Misc. services	6	1	7

Bearing in mind the considerably bigger small firm sector in other industrialized countries, it can be illuminating to examine some of their statistics. For this purpose the USA is a convenient example. In 1963 39% of all manufacturing employment in the USA was in small firms; in Britain the comparable proportion was only 31%. Since then the USA percentage has increased and the British figure has declined. (Japan is another prominent example of an increasing manufacturing employment share in small firms.) Data drawn from the Office of the Secretary of the US Treasury and the Office of Tax Analysis published in the *Harvard Business Review* (Jan/Feb 1979) shows a striking difference between earnings on assets between small, medium, and big firms in all five main economic sectors recorded.

Asset class ($'000)	Earnings per $ of assets				
	Manufacture	Services	Construction	Transport	Wholesale/ retail
< 25	0.49	2.42	0.85	0.42	0.49
500–999	0.22	0.13	0.19	0.15	0.20
10,000–25,000	0.14	0.09	0.09	0.10	0.12
> 100,000	0.10	0.07	0.05	0.05	0.09

These differences are bigger and more consistent than can be accounted for without recognition that the smaller the firm the more effective in monetary terms is its use of investment. This, combined with equally remarkable information which shows greater personal satisfaction and less incidence of disruption by employees, points to a clear superiority for American small firms in the overall use of human and financial resources.

From all the available evidence there would seem to be no grounds for fearing a general lowering of resource utilization if directions in future development seek to achieve the minimum efficient size rather than, as in the past, striving for ever increasing scale of operation. There is a fuller discussion of the prospects for small firms in Chapter 7.

The future of big firms

The minimum efficient size for a particular industrial or commercial activity will vary a good deal. Just as in nature the size of animals and plants covers a wide spectrum, with inter-dependence between each species, so in the business world there are appropriate minimum sizes for each interdependent function.

Some big firms will find a place in the future mix, but this does not mean that important changes may not occur in such organizations to make them more effective than they are at the present time. Some are already seeing the need to restructure themselves so that their individual operating units do not employ more than, say, 300 people.

Others, like some major paper manufacturers, are finding that they have in some instances increased plant capacities to such an extent that they have exceeded the maximum size for efficient operation. By looking at smaller plants they could well conclude that capacities of little more than one-tenth of the biggest existing ones are feasible efficient units which, when sited in a dispersed geographical pattern, can provide a hedge against increasing transport costs caused by the inevitable rise in energy prices. The human side of big firms may also undergo a major transformation, with the number of shop-floor workers continuing to decline as more productive plant and routine office work is taken over by data processing machines. With the development of microelectronics, there is an opportunity for many of the existing in-house overhead services to be carried out by individuals or small groups who provide the services under contract. Dr Gershuny, in his book *After Industrial Society*, shows that only about half the people currently employed in the 'services (tertiary) sector' of the economy provide services to individuals, while the rest provide services to organizations. As the human side of big firms changes, many more people are likely to appear as units in the 'service sector' statistics, but they will be concerned with serving organizations, including big companies, rather than serving individuals.

In effect, there is likely to be growing pressure within many big companies to discover ways in which, without totally dismembering themselves, they can gain the kind of advantages that smaller firms enjoy.

Manufacture, distribution, and sale

All businesses are concerned with the provision of goods or services to a particular market sector. However, many modern big businesses are not entirely integrated in the sense of processing raw materials and selling them through their own wholesaling and retailing divisions to the ultimate buyer. Very often the producing company sells its products to various 'middle-man' firms which in their turn store and transport to a network of 'reseller' companies that complete the distribution to the ultimate customer. As a broad generalization, economies of scale apply only to the production process. The bigger the scale of production (plant capacity), the bigger the area of distribution that is required to absorb the output. Hence distribution costs tend to increase as plant capacity increases, and this increase prevents all of the economy of scale appearing in the retail price of the product. If a wide range of consumer goods is selected an approximate typical breakdown of costs, including profit, for £1 of retail sales price would be:

Production cost	40p
Manufacturer's overhead	20p
Sales price to distributor	60p
Distributor's costs and overhead	25p
Retailer's cost and overhead	15p

Retail sales price £1.00

Manufacturer's overhead is seldom reduced by bigger plant capacities; indeed, overheads frequently increase with the scale of operations. Thus only the 40p production cost is beneficially influenced by economies of scale, and since between 50% and 70% of production cost is the cost of raw materials, which are realistically unaffected by the scale of operation, it is really only 20p or less of the retail price of £1.00 that can be influenced by plant size. Most modern industries have a view as to the optimum viable size of plant in that particular industry. Estimates have been made of the loss incurred by building plant only one-third the optimum size and in very few cases indeed does the production cost increase by more than 5%, or 2% on the retail price if no additional distribution cost is incurred. However, if a small-scale plant operation can eliminate the need for a 'middle-man' (distributor), a substantial part of the warehousing, handling costs, transportation, and overhead can be eliminated, giving a potential saving of the order of 10% on the retail sales price. Unfortunately, this is only possible with certain types of product and certain markets.

The obvious question, then, is 'why does large-scale production often result in lower-priced products in retail stores?' One reason is that big, powerful distributors and retailers are able to obtain sizeable discounts or rebates from manufacturers. As a result small distributors and retailers pay prices for goods which in effect contain an element of subsidy for their big competitors. Although it is slightly cheaper to distribute to big customers, the savings are very small compared with the discounts given, which may be as much as 25-30%. Another reason is also associated with price differentials which apply to small and big companies. As already mentioned, a manufacturer's production cost will typically consist of 50-70% purchased components, raw materials, packaging etc. A large manufacturing company may be able to use its purchasing power to obtain very big discounts which are not available to a small company. The combination of these two consequences of 'purchasing muscle' is a major cause of the difference in pricing that appears in Britain between the products of big and small producers, rather than technical economies of scale. If a company enjoys a 20% discount on all its purchases and these represent 50% of the price at which it sells its products, the discount is the equivalent of 10% of the product sales price. Since companies seldom have profits exceeding 5% of their sales revenues the purchasing discount accounts for more than the whole of the profit. Without these big discounts such companies would be operating at a loss; in other words they are subsidized by the discounts. Legal limiting of the level of discounts and rebates is required to ensure fair trading between big and small companies.

Production or services?

Most of this chapter has concerned the production and distribution of the goods part of the economy. Quantity of production is usually expressed in terms of a number of items per unit of time (e.g. 50,000 cars per annum). Although this is satisfactory for perishable items, such as foodstuffs, it may be misleading for durable goods. For example, houses built to last for 100 years cannot sensibly be equated with houses that are useless after 50 years. A truer measure of production for all durable goods would be expressed arithmetically as the numbers of items produced, multiplied by years of useful life. However, the useful life of articles is not only a function of intrinsic durability, but is to a considerable extent dependent upon careful use, maintenance, and repair. Thus an economy with a large maintenance and repair sector (manual services) will require only a comparatively small production capacity of durable goods. In this sense these services are an alternative to production, or are directly complementary to it, in the adding of value in durable goods.

Compared with production, maintenance and repair has been much less affected by labour-saving technology: in producing a certain quantity of added value in manual services considerably more labour is employed than is the case in a production operation. This does not mean that the average added value per employee in manual services is much less than it is in manufacturing industry. Even in the manufacturing companies of Japan, in which labour productivity is outstandingly high, the added value per employee is only fractionally above twice the average employee earnings. In British manufacturing companies it is typically about one-and-a-half times average employee earnings. In this country manual service workers have no difficulty in achieving that level of added value. Indeed, a productive worker is often able to achieve added value of more than double average earnings. Bearing in mind that there is little capital requiring compensation and only a minimum of overhead, manual services are in no way inferior to manufacture from the point of view of the amount of wealth created per worker. The nature of the work is mostly more skilled than that of shop-floor industrial workers and is considerably more varied. It is also work that is widely spread geographically and carried out in small units. In all these respects it is superior to manufacturing work. From the point of view of environmental and material or energy conservation it is much to be preferred.

It is frequently argued that extending useful life by extra servicing and reconditioning would have the disadvantage of delaying the spread of technical developments. There is little substance in this argument, since in the short term, say 10 years, technical improvements are mostly slight; those that are valuable can usually be incorporated during reconditioning at half-life.

Most of these points were well illustrated in a study carried out by Walter Stahel and Genevieve Reday-Mulvey at the Battelle Research Centre in Geneva. They were commissioned by the EEC in Brussels to explore the potential for substituting manpower for energy in the motor-car and building industries. They based their investigation on the French economy, where, at the time of the study in 1976, the average life of cars in the French population was approximately ten years. They envisaged doubling useful life by increasing the amount of servicing and repair and by introducing a reconditioning operation during the extended lifetime of each vehicle. Available statistics showed that the following quantities applied to the French car population in 1976 per year per 1000 vehicles:

Average life of cars: *10 years approx.*

Labour used in production
    (materials production plus vehicle manufacture)   14 man years
Labour used in maintenance, repair   20 man years
Total labour used   34 man years
Total energy consumption (tonnes/oil equivalent)   150
    (57% in materials production, 43% in manufacture)

The effect of doubling the average life of cars to 20 years is shown in the following estimated quantities:

Average life of cars: *20 years approx.*

Labour required in production   8 man years approx.
    (materials production plus vehicle manufacture)
Labour required for maintenance, repair and reconditioning
                                           45 man years approx.
Total labour used   53 man years approx.
Total energy required   87 tonnes oil equivalent

Over a period of 20 years 280 production man years plus 400 service man years are required with 10-year average life cars, making a total of 680 man years per 100 cars. Over the same period, with 20 years' average life, 165 production man years plus 900 service man years total 1065; that is 385 more than with 10-year average life vehicles made up from an increase of 500 in the service sector and a reduction of 115 in factory production work. Of course, if some of the production work took place in the factories of overseas car manufacturers (as is the case in Britain, where more than 50% of new cars are imported), the effect on local employment would be even greater.

It is important to consider the financial implications of such a change in average car life. By doubling the life there would be a saving equal to the purchase price of one car every 20 years to offset the extra cost of servicing and reconditioning. The saving of the cost of 1000 cars, at £4,000 per car, would be £4 million. Providing that the additional annual servicing cost per worker was not more than £10,000, there would be a net saving to motorists who adopted the achievement of longer car life by means of servicing and reconditioning. The availability of more and better-quality work, widely dispersed, would be supplemented by substantial savings in both energy and materials with an accompanying reduction in waste and pollution.

Unfortunately we have become accustomed to the mistaken idea that added value is created most effectively in production industries, and consequently that labour-intensive manual services are comparatively inefficient. For a car with an average life of 10 years, and a first cost of £4,000, the annual value of the 'production added value' is £400. Average earnings at the time of writing are about £3 per hour. Even if those earnings were £5 per hour, an extra 80 hours per year of servicing, repair, and reconditioning could be afforded to add one more year – valued at £400 – to the life of the vehicle. Substantially less than 80 hours per year would be needed to obtain that extension. By extending average life beyond current levels through more servicing, repair, and reconditioning, there is a more effective adding of monetary value than occurs in current high capital production methods. We must not make the mistake of assuming that by substituting labour-intensive services for capital-intensive production we would be taking a retrograde step in the effective and efficient use of resources. The opposite is true.

The above example illustrates what could beneficially be achieved for many durable products if the present 'use and throw away' system of consumerism were replaced by a conserving system. At a meeting of OECD-nation experts in Paris in 1980 it was agreed that a move towards a greater emphasis on maintenance, repair, reconditioning, re-use, and recycling offers the best prospect for an improvement in the quantity and the quality of manual employment in industrialized countries. This is a key factor in a move towards a sustainable and more humanly satisfying form of society.

The change of emphasis required will need considerable changes in the rules that regulate the economic game, because the existing rules have been deliberately designed to encourage the growth of production and consumption. The rule changes required will be discussed in Chapter 10. They will not easily be adopted since there are powerful vested interests among employers, trade unions, financial institutions, and professional bodies that can be expected to oppose them. However, the change of emphasis is not entirely dependent on rule changes. Already there is evidence of a movement in some parts of the economic system. Renovation of buildings is beginning to displace some of the wholesale 'demolish and rebuild' policy of the early post-war decades. Bottle banks and can recycling skips are evidence of a growing recycling movement; and there are increasing numbers of shops specializing in the resale of used articles, particularly clothing, either for charities such as Oxfam, or for private profit – the 'nearly new' shops. Signs of improvement in maintenance, repair, and reconditioning are less apparent; satisfactory services are frequently hard to find and are of poor quality and absurdly expensive. Nevertheless, since the mid-1970s when average car life in Britain was about 10.7 years it has slowly been increasing until at the time of writing it is in the region of 13 years – still far short of the Swedish level, now more than 17 years. (Between 1966 and 1974 average car life in Britain decreased from 15 years to 10.7 years while in Sweden it increased from 10 years to nearly 15 years. It is thought that a severe form of MOT-type testing played a considerable part in the Swedish improvement.)

Checkpoint

There is so much confusion and misunderstanding concerning productivity that it will be worthwhile to summarize some of the main points of this chapter.

1. In many industries or businesses it is not only high capital/labour ratios that are highly productive: there is a range of mixes of capital and labour which can be equally cost effective.
2. It is not true that higher overall levels of productivity can only be achieved by increasing the ratio of capital to labour. For a long time, while output per man has increased, output per unit

of investment has been decreasing. The most efficient use of manpower and money is achieved when the amount of each that is required to add value is reduced.

3. Optimum systems of production and distribution are not always obtained simply by maximizing scale economies of production and then designing an efficient system to distribute the output. Frequently much smaller production units, with minimal associated distribution costs, produce lower total costs for the complete system.

4. It is a fallacy that high rates of 'adding value' per man can only be achieved in capital-intensive production. Much of the work done, with very little capital investment, in maintenance, repair, reconditioning, re-use, and recycling has rates of 'adding value' as high as or higher than a great deal of production.

There is no intrinsic merit in seeking high labour intensiveness on the one hand or high capital intensiveness on the other. The very essence of appropriateness is to find that particular set of circumstances which will result in an effective and efficient overall use of resources with maximum opportunity for human fulfilment and minimum environmental damage. The particular circumstances that prevail in Britain suggest that this can often best be achieved in small or medium-sized units with modest levels of capital investment.

CHAPTER 3
Forces at work

The difference between work and employment

In our review of the changes that have taken place since the Second World War mention was made of the transformation that has taken place in patterns of employment and the loss of job satisfaction that many people have suffered. In seeking to find a more satisfactory alternative to the present system it is important to consider improvements that can be made in working life.

Among the changes that have occurred in the last few decades were significant movements of activities from domestic work to industry or to paid-for services and vice versa: traditional food and drink preparation has been partially industrialized, and household decoration is frequently carried out as a DIY activity. Satisfactions and dissatisfactions come from all forms of work, whether paid or unpaid. In considering future alternatives we need to consider work in its several forms – paid employment, domestic, and voluntary – particularly as most people are likely to spend a smaller proportion of their lives in paid employment.

In the past two hundred years we have become a very money-dependent society. This is partly because many domestic tasks, or work and goods that were bartered or exchanged, have now become part of the money economy. Paid employment has become an extremely important factor in most people's lives. It is sometimes argued that the lack of a job would present no real problem if being unemployed did not usually mean losing a considerable amount of income. If we are to consider ways in which we might conduct economic affairs differently, we need to know how important it is for every able-bodied person to have a job. Would people be satisfied without a job providing they suffered no financial loss? Or, looked at another way, what does paid employment provide in modern society in addition to income?

A considerable amount of research in recent years provides answers to this question. For most people the most important thing that a job provides is a sense of identity, purpose, and status, by making a regular daily demand upon people to which they are obliged to respond. There has been some suggestion that this phenomenon is merely a byproduct of the work ethic which has been prevalent for the past two hundred years. The fact that many family names such as Smith, Baker, Miller, Carpenter have ancient historical origins suggests that a person's occupation has long been a vital social symbol. It is not surprising, therefore, that people who are without a job often feel they lack an identity: they are without purpose, motivation, or prospects. Social contacts are the second most valued benefits a job provides, often felt to be the most important relationships in life outside the immediate family. Their value owes much to the orderly structure and the common interests and purpose provided by the organizational context in which they occur: generally speaking, people know where they stand and what is expected of them. A comfortable life depends to a considerable extent upon the existence of generally accepted codes of behaviour which are not too onerous, and also upon regular rhythms of activity. A job provides an invaluable daily structure of time and discipline in which roles are fairly clearly defined and understood. Many people who become unemployed experience a sense of bereavement from the loss of friends, colleagues, and facilities and of isolation at their exclusion from what they feel is the mainstream of life. Even the activity involved in paid employment is felt to have a special quality because it is purposeful and reward-associated, providing the kind of added satisfaction experienced when a game is played for a prize. Lastly, the money obtained through employment is felt to be a special kind of money, partly because it is earned but also because it has been in one way or another negotiated. Memories of a first pay-packet embody this feeling at its height. Money received gratuitously is felt to lack the satisfaction of money negotiated and earned, and the recipients are left feeling dependent.

This catalogue of benefits gained from paid employment is so rich in human satisfactions that any acceptable alternative form of economic development must surely provide opportunities for all to enjoy them throughout adult life. Although all kinds of work can provide satisfaction, they are incomplete when the work is carried out by people who are classed as unemployed.

This does not mean that other forms of work are of little value and do not have an important contribution to make to wealth creation and human satisfaction. The increasing proportion of non-manual jobs leaves many people lacking opportunities to exercise their potential for practical skills. Some jobs provide no opportunity for direct person-to-person service. Both of these shortcomings can often be met by domestic or voluntary work. With a smaller portion of the day being spent at the factory, shop, or office these forms of work could become more important in the future.

The research work of recent years provides strong confirmation for the belief that 'work is made for man' and not man for work, which is the view taken by those who see people as mere factors of production.

It reinforces Article 23 of the United Nations Universal Declaration of Human Rights (1948) which states that 'everyone has the right to work, to have free choice of employment, to just and favourable conditions of work and to protection against unemployment.' Furthermore, this belief makes clear the need of paid employment that exists for most people, as well as of domestic and voluntary work. The aim of future development must not be merely a job for everyone, regardless of its nature, which occupies so much time that there is little opportunity for other forms of work. Instead we should look for ways of conducting our affairs so that they provide a mixture of satisfying paid employment for all and sufficient time and resources for people to enjoy the benefits of voluntary work and work at home.

Popular fallacies about employment and unemployment

With more than three million people without jobs, and some forecasts of worse to come, this aim may seem to be absurdly idealistic. We need first to clear our minds of a great many myths and mistaken ideas associated with employment and unemployment. There are too many to be catalogued exhaustively, but the most important ones are:

1. Economic growth is necessary to reduce unemployment. If this were true the United States would have not more but less unemployment than Japan, which has less than the American GOP. Unemployment in Britain would be less now than it was in 1960 instead of being almost ten times as high. And yet there are still demands for economic growth as a means of reducing unemployment.

2. Modern capital-intensive technology, it is often claimed, is the real culprit, despite the fact that more people are in employment now than in the past.

3. Unrestricted imports of goods from countries with low labour costs is claimed to be the main cause of unemployment in British manufacturing industries. In fact a much higher proportion of imports comes from countries where labour costs are higher than they are in Britain.

4. Even though we are dependent on imports for more than a quarter of all the manufactured goods that we need, and for 40% of our food, it is said that lack of demand in the economy is a cause of unemployment.

5. Despite the fact that some countries with considerably higher wage levels succesfully operate manufacturing activities with a high labour content, it is sometimes claimed in Britain that we should

abandon such activities to Third World countries and concentrate on new high technology / low labour content activities in order to ensure full employment here.

To set against these misleading ideas there are some truths which receive far too little attention:

1. There is an almost unlimited amount of work to be done. It is paradoxical that even though there is a high level of unemployment much work remains undone. It seems likely that the work gap will be filled partly through paid employment and partly through voluntary and domestic activity.

2. There is a limit to material needs in this country and also to some of the world's natural resources. With less work required for any given level of production, it is no use expecting net increases in employment in the manufacturing side of the British economy except where substitution for imports can be achieved. Although Britain has a much lower share of international trade than in the past, it still has more than a fair share of the trade for manufactured goods. There is no reason why that exceptional historical position should be maintained, and it is virtually certain that the share will continue to fall to a level in keeping with the size of the British economy. As saturation of the domestic market is approached, the future scope both for wealth creation and industrial employment will depend largely on whether or not total international trade expands. Although there is much talk of an expansion as the Third World develops, most evidence points to a form of development in all countries which is less dependent on international trade. With very limited scope for long-term expansion in the material side of the British economy, and a large part of the labour intensive side being in the public services (which are dependent upon limited tax revenue derived from the private sector), it would appear that most job opportunities will need to be found in the private service sector. Many satisfying high value added jobs can be created in repair, reconditioning, re-using, and recycling services – the 4Rs. It is in these manual services that the vast number of manual workers who have become redundant to the needs of manufacturing industry and agriculture will be able to earn their living once again. It is also from an expansion of these private sector services that additional tax revenues can be developed to provide for improved public sector services.

3. The level of unemployment depends upon the rate at which existing jobs disappear and the rate at which new jobs are created. Current

high levels have occurred because the rate of new job creation has not been sufficient to re-employ the rapid and massive exodus of people, from manufacturing industry particularly. Those countries which have suffered less during the recession are those which were more internationally competitive, had less overmanning in their industrial companies, and which had historically a much better record of new job creation by means of a flourishing small firms sector. Although there has been some improvement in the rate of new job creation in the British small firms sector the measures applied here still lag far behind most other countries. In particular almost nothing has been done to promote the development of manual services, where the best employment prospects exist. Instead, vast resources continue to be concentrated on wealth creation with little associated employment, much of it, regrettably, with limited success, so that even the indirect employment that arises as a byproduct of wealth creation has been disappointing. Ten years of experience since unemployment began to increase rapidly provides sufficient evidence that market forces alone do not bring about the desired change. Governments must change the rules of the economic game quite radically if a change of direction is to be effected.

4.   Even if the necessary stimulus were to be given to create all the new jobs that are desirable, and even if it was completely successful, there can be no guarantee that full employment of 48 hours a week, 48 weeks a year and 48 years a lifetime would return. It seems highly improbable. It also appears to be undesirable, because many who are currently employed in that particular work pattern suffer considerable stress and job-induced sickness. One of the saddest aspects of the present situation is that while many are deprived of a job, many more with jobs have too much work to do. In the final analysis full employment can only exist when the total amount of work that needs to be done in the formal money economy is equitably distributed among the whole working population. Whatever else may be done to stimulate the generation of wealth-creating jobs, full employment will only return when measures have been introduced to facilitate work-sharing. One of the most important happenings of the past century was the widespread acceptance of the idea that a just society requires positive steps to be taken to aim at fair shares of income and wealth for all. Now that we can better understand how much more than money is provided by a job, it is surely clear that it is at least as important for paid employment to be shared fairly as it is for income and wealth.

Sharing work

Talk of work-sharing is usually met with scepticism, possibly because it is thought that income reduction is implied. A considerable degree of work-sharing occurred during the postwar years when there was full employment (less than 500,000 unemployed) and incomes steadily rose. Had that sharing not taken place, and if weekly hours were still 55 to 65 hours and annual holidays only a week, current unemployment levels would be very much higher than they are currently – about six million rather than three million. Work-sharing was effectively the result of trade union pressure for improved labour productivity to be rewarded with shorter working hours as well as improved rates of pay. It is more important than ever, now that high rates of unemployment have returned, that this process should continue, if anything with more emphasis on shorter working hours. If increased productivity is used to shorten working hours, rather than to increase rates of pay, more room will be left for the attainment and maintenance of full employment, albeit on a shorter working week.

Work-sharing is a reallocation of working time, reducing the working hours of the employed in order to spread employment opportunities amongst those without a job. Despite the fact that other industrialized countries have achieved greater affluence with considerably shorter working hours, work-sharing is frequently opposed in Britain on the assumption that it will increase unit costs and damage competitiveness. In the Department of Employment Research Paper No. 38, M. White reported that 64% of the firms which had reduced working time had experienced no reduction in overall effectiveness. In such firms there would be no increase in unit costs even if hourly wage rates were raised fully to maintain weekly gross pay. Conversely, it has been estimated that unemployment costs each person in a job £700 per year on average in higher taxes. Reducing unemployment by work-sharing would give scope for reducing some of this tax burden on employed people. If this were done, and wage rates were adjusted to maintain after-tax income, which determines living standards, unit costs could be reduced, thereby improving competitiveness.

The presumption that shorter working hours increase unit costs depends on the twin assumptions that output would be proportionately reduced and wage rates would necessarily be proportionately increased to maintain a full weekly wage packet. Both assumptions can be challenged from practical experience. It is clear, however, that if there is to be progress in work-sharing practices there is a need for close co-operation between management, trade unions, and national government. In the face of scepticism and opposition it is most likely to be achieved if there is recognition of a mutuality of interests at a time of rapid technological change which is liable to reduce labour requirements. In these circumstances the benefits of new technology can be effectively shared out within society by means of work-sharing, rather than going solely to a dwindling remnant of a working elite and the owners of capital.

Probably the most urgent way in which progess is required is in the reduction of overtime. Not only is British labour productivity poor by comparison with most European countries, the average hours worked per week is very high. These factors are not unconnected. Overtime working has become institutionalized and is widely worked on a regular routine basis on demand by the workforce in order to supplement take-home pay. This has the effect, described by C. Northcote Parkinson (of *Parkinson's Law*) of work being spread to fill the time available – hence low productivity. Neither the workforce nor the management needs to fear that output or incomes would fall if overtime working was reduced. If any proof is needed it was provided by the 'three-day week' experience of 1974, when almost as much was produced in three days as was previously being turned out in five. Although a radical reform of overtime working in Britain might not directly contribute to a substantial increase in jobs, it is a very necessary part of the process of working hours being reduced in step with productivity increases. Some European countries restrict the number of hours of overtime that may be worked by any employee during a year. In Sweden overtime is restricted to 150 hours per year; in West Germany a maximum of two hours per day is stipulated. This ensures that overtime can only be used to meet exceptional circumstances. As a result the average working week in Sweden is 36.2 hours and in West Germany 40.1 hours, compared with about 48 hours in Britain. Such a restriction in Britain would no doubt meet with strong resistance from both management and trade unions, but there is hope that the EEC may introduce legislation that would make it obligatory in all member countries.

Not only has work-sharing proceeded imperceptibly through many years by the gradual reduction of working hours for full-time workers, but there has also been a steady increase in the proportion of part-time workers. This has to a considerable extent paralleled the increase in female employment and reflects the fact that considerable numbers of people prefer a part-time job to one that demands a full week's work. Although in some instances there has been initial reluctance on the part of managements to accept part-time workers, many have found that they can offer marked advantages in a great many types of work, not least an increase in productivity.

The subject of job-sharing, which can take many different forms, has been pursued much more actively on the European continent than in Britain. In many instances employers have been agreeably surprised by productivity improvements that have accompanied the division of a job between two people. Already in Britain 20% of workers are employed for only 20 hours a week and it is believed that many more would opt for part-time work if they were given the choice. In the House of Lords Select Committee on Unemployment a possible method of encouraging 'pairing' of jobs was suggested, in which one member of the pair comes off the unemployment register. The suggestion was that the government might offer such part-time 'paired' employees full-time National Insurance benefits on the basis of half-time contributions. At the same time the NI cost to employers would be the same for the pair as for a single full-time employee. The Committee calculated that there would be an overall net benefit to the Exchequer. There are two particular circumstances in which such pairing is exceptionally advantageous: firstly, when a job is split 50:50 between an elderly worker desiring a less stressful job and an unemployed young worker; secondly, when a job is divided between two young trainees. GEC Telecommunications Ltd in Coventry pioneered a scheme in which one job and one wage is split between trainees. There are two advantages to employers, the chance to build up a bigger pool of trained workers and a reduction in absenteeism. Widespread use of this scheme could be a valuable way of spreading skill training and employment among school-leavers.

Sharing either work or income involves some sacrifice on the part of the donor. If work is not shared and high levels of unemployment persist, the minority of people unemployed make a very big sacrifice in order that the majority are unaffected. Justice demands that all should share in whatever sacrifice has to be made, and this involves some redistribution of income from those currently at work to those who are to be brought into employment. Faced with the choice between taxes being raised to give unemployed people incomes comparable with those earned in employment and the sharing out of work, with reduced pre-tax earnings but no increase in tax, would not most people choose the latter alternative with its increase in leisure time for all? It is this choice that must be made in any country that lays claim to freedom with justice for all its people. When it is recalled that a paid job provides people with a great deal more than a source of income, work-sharing can be seen to be a great deal more than mere redistribution of income.

Redistribution of paid employment is an imperative if justice is to be done, and there is no prospect that it can be done painlessly. It is not surprising that governments introduce schemes which attempt to avoid the unpopularity of painful measures. It is equally unsurprising that such schemes, which depend upon individual voluntary acts of self-sacrifice, meet with little success. For example, a part-time job release scheme, allowing people nearing retirement to work part-time and receive a supplementary allowance provided an unemployed person was recruited, attracted only 26 participants throughout Britain in the first six months. Another job-splitting scheme, which it was estimated might produce an additional 62,500 jobs, attracted only 774 people in the first year, even though a grant of £750 is paid to the employer who splits the job. As a signatory of the Declaration of Human Rights, it is surely incumbent upon Britain to take whatever steps may be required to ensure that all who want a job should have one as a matter of basic human right.

The costs of unemployment and of employment

So far in this chapter we have considered the cost in human terms, as well as income loss, that employees incur when they become unemployed. Now we should consider the considerable cost of unemployment to the whole population, and compare it with the cost of providing people with a job.

In 1982 the average annual earnings of all men and women in employment was about £5,800. The average value of output per employee was about £11,500 per annum: wealth was being created at a rate roughly equal to twice average earnings. Had the 3-4 million unemployed people been creating wealth at the same average rate as those in employment an extra £34 to £35 billion would have been added to the gross domestic product (GDP).

The financial cost of unemployment is usually published, not in terms of lost wealth creation, but only in terms of payments that are made or revenues lost out of the public purse. Even at Britain's low levels of unemployment benefit – considerably lower as a proportion of average earnings than in all comparable European countries – in 1984 the Treasury estimated a cost per unemployed person of approximately £5,000 per annum (compared with about £1,500 per annum unemployment benefit), which was only £800 below average earnings of employed people. The estimate is a fairly complicated one if all the direct and indirect elements of cost are to be included, and different judgements may be made about what to include or leave out. Although the Treasury consider the cost to the Exchequer is somewhat less than average earnings, others believe that it may be even greater. Other bodies have made estimates greater than the Treasury figure: the Institute of Fiscal Studies produced a figure of £7,500. This immediately raises the question of whether it might not actually be cheaper for many more to be employed in public services than to remain unemployed. The costs of such employment would have no adverse effect on international competitiveness so long as they did not exceed the alternative cost of unemployment. There would be no question of overmanning, since there is a need for more public services that this strategy could usefully provide.

Not all the 3-4 million unemployed could be provided with public service jobs, so the cost of creating additional jobs in the private sector needs to be considered next. The impression gained by the public from media announcements is that very large capital sums are necessary for every job created in private industry. Projects involving investments of hundreds of millions of pounds are said to offer employment for only a few hundred people: for example, a new car plant or large chemical project may require an investment of between £100,000 and £250,000 for each job created. Frequently, in order to attract such projects to the United Kingdom in competition with other European countries, the British government has been involved in paying heavy capital grants equivalent to tens of thousands of pounds per job.

These examples, which received so much publicity, are quite unrepresentative of the average capital investment in a single job in Britain. The averages of about £20,000 per job in manufacturing industry and £6,000 to £8,000 in the construction and service industries are composed of a wide spectrum of investment costs for different activities. It is also true that within a single type of activity – printing, for example, or brewing – there can be a fairly wide spread. When a new business is being set up the capital investment can often be minimized by the acquisition of second-hand plant in good condition at favourable prices; premises can be rented instead of purchased; and some equipment may also be rented or leased. It is not unusual, therefore, for jobs to be generated with capital investment by an employer of no more than about £10,000 per job in manufacturing and for less than £5,000 per job in service activities. Given that part of the capital – working capital – can usually be borrowed from a bank, providing adequate security can be offered, it is fair to say that on average a job can still be produced in Britain at a cost to a private investor of no more than about two years' average earnings.

The cost of creating work-places (jobs) is a key factor in any consideration of wealth (added value) and employment generation. In his writings about Third World development, Schumacher emphasized the need to provide technologies capable of adding value at a higher rate than traditional technologies at costs per work-place that are within the financial reach of ordinary working people – two to two-and-a-half times average annual earnings was suggested. Such moderately priced technologies – which he called Intermediate Technologies – are capable of providing large numbers of jobs with higher earning potential. In contrast, expensive advanced technologies imported into those countries from highly developed industrial economies provided few jobs and simply created a small, rich elite, leaving the mass population in poverty. Countries such as Britain are showing signs of a similar phenomenon developing, as an increasing proportion of industrial investment is channelled into highly capital intensive mammoth projects which provide good incomes for their few employees, leaving a steadily increasing number of people without employment as an under-class. The wealth-creating potential of a nation, or a town or village, can only be attained when all the work-force are employed. Since the average cost of generating work-places in Britain – as in the Third World – is no more than about two years' average earnings, there is no financial barrier preventing the potential from being achieved. It is surely an absurdity for governments to be paying in unemployment benefits and other costs something like average earnings, with a loss of double that amount in national wealth creation, when an investment of only double average earnings is required to provide the jobs that are needed to abolish the evil of unemployment for men and women and their families.

While it remains true, therefore, that in the final analysis there can only be full employment when the paid jobs available are equitably shared among the working population, it should not be assumed that there is only the present limited number of jobs to be shared out. There is a potential in both public and private sectors of the economy to generate many more than the twenty million jobs currently in existence. Although many are likely to offer conventional kinds of work, some will be new types of job arising out of the many significant changes that are certain to occur during the coming decades. There were not many energy consultants before energy became expensive, and not many sports or leisure centre assistants until such places sprang up all round the country.

Stimulating employment development

In looking for ways of stimulating employment development that are appropriate to present and expected future circumstances, let us first consider the obstacles which are currently impeding the creation and expansion of small and medium-sized firms and holding the manual services at their existing low levels. These obstacles can be broadly categorized under the three headings of physical, financial, and environmental.

One of the commonest obstacles to new firm creation, particularly in manufacturing or manual services, is the absence of low-cost small workshop facilities, between, say, 200 and 700 sq.ft of floor area. Indeed, until very recently there has been very little available of less than 1,500 sq.ft. The burden of paying rent and rates, heating and lighting, for up to five times as much space as is required is a real inhibition, especially if a long-term lease is a condition. There may be equally serious obstacles in obtaining adequately trained and experienced employees, if the new business happens to require skills unusual in the area. Such a problem faced a company establishing a fur product factory in a steel closure area. Ex-steel workers required several months of costly re-training before they became productive. A very large proportion of industrial skill training has traditionally been carried out by major companies and has been of great benefit to all sizes of firm. With the increase in shop-floor labour productivity and the associated de-skilling of many jobs in major companies, serious shortages of skilled workers have developed in the absence of alternative training schemes. Shortage of skilled people is equally an obstacle in manual services. Here the burden of obligations on many independent tradesmen in employing apprentices or mates has increased to such an extent that serious shortages are now developing. Action by both public and private bodies, sometimes independently but often in collaboration, is required to provide suitably sized facilities to meet a wide variety of needs and to replace the traditional training schemes which no longer meet the requirements of small firms who themselves have neither time, money, nor facilities.

As the UK economy has become increasingly concentrated and centralized into a small number of big firms in both manufacturing and marketing, for certain types of small manufacturer obstacles have arisen in marketing. For example, finding outlets for a small independent brewery's products can be difficult in an area where most public houses are tied to one of the major brewing companies. Experience has shown that some easement of the marketing problem can be obtained if special facilities (e.g. low-cost covered markets) are created to meet the needs of certain types of manufactured goods; but beer is not, unfortunately one of them.

Another physical obstacle to small enterprise development is the high proportion of total research and development investment directed towards large scale industrial application. There is nothing comparable to the research and development organizations which exist in some other countries, such as Denmark, and make a valuable contribution to small-scale industrial development and innovation.

Financial obstacles are considerable, although they are not so crucial as is often thought. There is much disagreement about the effect of increasing manpower costs on employment. Those who claim that there is little evidence that the additional cost incurred in employing extra people inhibits firms from taking people on appear to rely mostly on large or medium-sized employers for their information. The inhibitions are greatest at the smallest end of the spectrum where the addition of one extra person can impose an intolerable burden; but, because of the large number of such firms, failure to increase employee numbers prevents a large number of jobs being created. It is by no means only the cost of wages that is put into the balance: incidental costs can easily double the wage cost burden.

Although there are very real obstacles to major business expansion in both small and large firms because of the exceptionally high rate of return that must be earned on normal loan finance, the problem with new enterprises is not so much an absolute lack of sources of suitable finance as the difficulty of gaining access to it. It is sometimes not understood that the short-term loan schemes that are available to businesses in Britain require earning rates on new investment a great deal higher than the going interest rate if the loan is to be paid off on time. When interest rates are about 20%, payments to the bank may well be nearly 60% per annum on a three-year loan. The business expansion problem can only be overcome by providing medium-term loans at privileged rates of interest for small firms, such as exist in most other industrial countries. The problem for new ventures can be overcome by a revival of venture capital schemes that are attractive to both borrower and lender, as well as by low-cost professional assistance to entrepreneurs from people who know how to gain access to appropriate sources of finance. In the long term, solutions to problems would be greatly facilitated if a much more diverse and dispersed form of banking took the place of the present monolithic and centralized system.

The environmental obstacles are too numerous to describe in detail. For any intending entrepreneur to overcome them can only be described in terms of penetrating a cunningly contrived system of minefields and barbed wire entanglements. Mountains of legislation have been enacted in the past decades, almost all of which is specifically designed to be applicable to big companies with large professional staff departments. Unfortunately it is applied indiscriminately to all enterprises, no matter how small. The removal of most of these obstacles by changes in legislation specific to small firms is vitally important and will be discussed in Chapter 8, 'The rules of the game'. When the new rules are drafted the unique condition of 'infant' small firms needs to be recognized and differentiated from 'mature' small firms. From conception to birth, and for the first three or four years of its life, a new small firm is an exceptionally vulnerable organism which requires a very special nursery environment if 'infant mortality' is not to remain at unacceptable levels. Until such time as the rules are changed there is much that can be done with experienced professional help to guide new entrants into the business world and over the obstacles that confront them. Local organizations which provide this kind of help, and which perform some of the roles mentioned with regard to physical and financial obstacles, are described in Chapter 9, 'Taking the initiative'.

But there is more to be done in stimulating employment than removing existing obstacles. Starting with conservation, re-use, and recycling, there is plenty of scope for positive measures which would not be inflationary. In 1980 there was an enormous quantity of material available in the UK for recycling – about five million tonnes of paper, two million tonnes of plastics and rubber, two million tonnes of glass, and over nine million tonnes of ferrous and non-ferrous metals. Despite the publicity given in the last ten years to the opportunities that exist for recycling, recovery rates are probably lower now than they were a few years ago; and since 1950 Britain has had a poor record compared with some other countries and with earlier times. For example, a substantial industry existed in west Yorkshire in the production of low-cost woollen goods from reclaimed wool. Now most used woollen materials are exported to Italy, France, Spain, Portugal, and Holland where they are reprocessed in small firms.

The most neglected area of recycling is in domestic waste which amounts to some 20 million tonnes per annum. When the waste is mixed there is a problem of separation, but experience indicates that most householders are willing to sort their domestic waste if they are provided with the means of doing so. Many attempts to recycle domestic newsprint commercially have failed because of large fluctuations in demand and price. These problems can be at least partially overcome if collection and reprocessing operations are integrated or, alternatively, if the public goodwill that undoubtedly exists – witness the success of 'bottle banks' – is harnessed to bring waste material to a collecting point. There have been many impressive initiatives in recycling in the United States which could usefully be followed in Britain. However, the full potential for waste recycling is unlikely to be achieved unless measures are introduced by government's giving preferential treatment to recycled materials.

There has been growing concern in Britain about the pollution created by agriculture. Straw burning is only one aspect of the problem. Six million tonnes of straw are wasted each year which has a potential value of £700 million as chemical feed stock. Collaborative research and development by Birmingham University and ICI has produced a process which is a breakthrough in the conversion of straw to chemical feedstock. When in a few years' time this conversion process is in full production, if only half the feedstock value is absorbed in wages about 70,000 jobs would be created, most in gathering and transporting the straw.

Recycling materials is very important but is the last phase of the 4Rs conserving process – repair, reconditioning, re-use, and recycling. Recycling itself could add a great many valuable jobs, but repair, reconditioning, and re-use could produce many more jobs of good quality. The example of extending car life described in Chapter 2 referred only to the additional jobs that could be generated in car reconditioning. It did not include the extra jobs that are possible if individual components are reconditioned instead of replaced by new spare parts. In the last ten years substantial businesses have been developed in the United States in car component reconditioning and a similar development is beginning to take place in Britain. If it reaches its full potential, covering all kinds of machine components, many new jobs will be created and a great deal of valuable material and energy will be saved.

To obtain these employment benefits innovation in organization is necessary. It cannot develop out of the existing servicing operations, which are closely linked with the manufacturers of components and finished products. An attractive system could be developed in the form of ethical franchises, each one of which would specialize in a particular make of product. Hitherto most franchises have occurred in retail shops such as Kentucky Fried Chicken, restaurants such as Little Chef, or services such as office cleaning. Few are concerned with repair or reconditioning work, but this is also the type of work and the kind of business which lends itself to franchising because it is replicable across the country.

Franchising can also apply to some kinds of manufacturing business. One of the most significant developments in recent years in production technology is the massive reduction that has taken place in the minimum economic production batch size. Consequently the total nationwide requirement of such products can be achieved in a large number of small plants spread around the country. The feasibility of efficient production on a comparatively small batch scale adds greatly to the potential for product innovation, which to a considerable extent is the key to increasing manufacturing wealth creation and employment.

This relationship was well described by David Foster in his book *Innovation and Employment.*

Another organizational variation which could play a useful part in the development of more good quality employment on domestic services is the concept of 'community business'. This type of service work is by its nature suited to individual craftsmen operating from their homes. As it currently operates it is frequently very unsatisfactory from both the customer's and the worker's point of view. Customers have no guarantee that the people they employ are sufficiently skilled; and they can easily find that the service is unreliable and unnecessarily expensive. The craftsmen have to perform the tasks of managing a small business, for which they are seldom properly equipped. A 'community business', set up on a non-profit basis, would receive requests from householders, plan and monitor the work and the charges of a carefully selected group of skilled craftsmen, and manage the financial and organizational side of the business.

An important part of domestic service is neglected because householders are too poor to be able to afford to call in help and unable to undertake the work themselves, perhaps because of age or disability. As the population of elderly people increases, so does the need for services at non-commercial rates. Part of the function of a 'community business' can be to organize part-commercial, part-voluntary work to meet these needs.

Other forms of community-based innovations can be applied to foster the growth of wealth creating employment. Some examples are described in Chapter 9, 'Taking the initiative'.

Much anxiety is expressed concerning the so-called 'black economy'. It is looked upon with disfavour because it is hidden from the tax gatherers; wealth is being created and the State is denied its 40% plus share of that added value. When consideration is being given to the elimination of such activities, it tends to be forgotten that a significant proportion of businesses which are now part of the formal economy had their origins in the 'black economy'. Of course not all 'black economy' operators end up as legitimate tax-paying businesses; but it is important for embryonic enterprises to have several years' tax-free holiday if they are to have a chance of becoming viable commercial taxable businesses. Rather than trying to drive new initiatives out of existence, it would in the long run pay to legitimize them with time-limited tax holidays. This is just one of a number of legislative measures which would discriminate in favour of infant enterprises.

CHAPTER 4

Technology choices

Who determines the technology?

There is no simple answer to the question of how technology choices are made and who is responsible for making them. A tiny minority of very rich people may be able to influence the technologies that are used in their own lives. For example, they may insist that regardless of cost all their food must be produced by labour-intensive, traditional methods and be prepared in their own kitchens with implements and equipment specially designed to meet their needs. Their household furniture and their clothes may be individually designed and hand crafted. Even their car may be a 'special'. But even the richest people will not be completely in control of the technologies in common use. For instance, the metals that they use, and many of the other materials and fuels, will all be supplied from the common sources of these commodities. Items such as nails, screws, nuts and bolts, pipes, wire, valves, are invariably obtained from the common pool. Even though they may generate their own power and pump their own water, the machinery will not differ from what is in common use.

At the other end of the spectrum of wealth there is very little choice indeed of the technologies that are used to provide people's needs. Apart from a few things that they may grow or make themselves, everything comes from the sources producing the cheapest, standardized, mass produced items. If they have some special needs which cannot be satisfied by standard products, then most likely they will have to go without. As wealth increases, so does the scope of individuals in matters of technology choice, but for 90% of the population there is in modern society little chance of avoiding mass produced objects.

In both products and manufacturing processes, the process of technological choice is highly complex. As more and more products are produced and distributed by fewer and fewer companies the crucial choices are concentrated among a decreasing number of people who, by the nature and size of the big organizations they serve, are increasingly constrained in the choices that they make. Big companies are only interested in products which can produce a high volume of sales in the shortest possible time; companies which manufacture production plant and machinery consequently concentrate on developing high volume units. As a result little attention is paid to small volume production,

US company assets ($'000s)	$ earnings per $ of asset (1972)				
	Manufacturing	Services	Construction	Transport	Wholesale/retail
< 25	0.49	2.42	0.85	0.42	0.49
500–999	0.22	0.13	0.19	0.15	0.20
> 100,000	0.10	0.07	0.05	0.05	0.09

Source: Harvard Business Review, Jan/Feb 1979

particularly in a country like Britain, where the small firms sector has declined to a far greater extent that in other advanced industrial countries and uniformity rules.

Another major influence on the choice of technology derives from the overriding interest among most big companies in growth of their capital investment, even though this might often result in a reduction in the rate of return on each pound invested. (Comparison chart of investment earnings for companies of different size – US data). This inbuilt tendency of big companies is reinforced by government measures which provide various forms of capital investment subsidy and place a disproportionately high level of taxation on labour costs compared with corporate taxes on investment earnings. The effect is to minimize the human factor in the production process, and increasingly it excludes all but the richest and the most talented and sophisticated people from technology choice. If it is admitted that changes in technologies can have important influences on people's lives, this trend towards restricted centres of choice must be considered undemocratic.

The main defence of this system is that, while not perfect in all respects, the benefits it conveys to the mass of people outweigh any disadvantages. It is argued that so long as there is a competitive market people will not be forced to buy things they do not want for lack of an alternative. If a very limited and short term view is taken that argument carries weight; but if the view is broadened and lengthened it becomes more questionable. Motor-cars were in widespread use for nearly half a century before legislation had to be introduced to increase safety and limit exhaust pollution. It seems unlikely that product durability would have had such a low priority if there had been less emphasis on glamorous features and high performance. It was not a matter of motorists' choice or a consideration of public interest that reduced the average car life in Britain from 15 years to 10 years between 1966 and 1974. It was not the average housewife that asked for traditional British cheeses to be replaced by inferior simulations of their ancestors; nor was it the 'rolling English drunkard' that prompted the breweries to substitute a carbonated keg malt beverage for traditional ale – until the Campaign for Real Ale caused a rapid rethink. And it is presumably long shelf-life qualities that have produced the indigestible British white loaf, rather than a keen desire on the part of citizens to consume limp slices of a foam-plastic-like substance.

The size factor – efficiencies and inefficiencies of scale

While it is certainly true that the mass of people have greatly benefited from the existing system and the way that technological choices are made, it is also true that the price paid is becoming more and more intolerable when the sole criterion of choice is the bottom line in a company's profit and loss account. With such a system only those tech-nologies which are of interest to big firms and are capable of producing a higher profit will be used, and all other options will be closed. For example, a big company developed a useful process for avoiding the generation of obnoxious smells in pig farms. However, nobody has been able to find a way in which the process could be commercialized and so its benefits are denied to people who suffer daily from smelly pig farms. Nearly thirty years ago a young metallurgist in a major indus-trial company patented a process which greatly increased the strength of light steel components, facilitating a considerable reduction in the amount of metal used. The company saw no use for it and it was shelved. For many years the benefits that such a process offers have been denied because it did not satisfy the sole criterion of choice: quick profits. This is not an isolated case: most big firms shelve potentially valuable developments because they do not fit their particular business interests, and few are enlightened enough to make such inventions available to smaller firms to exploit.

The concentration of technical talent in major companies is reflected in big government institutions such as Harwell and Farnborough, and the way in which governments regulate taxation and capital grants also reinforces big technology. Governments also have major influence on technology choices in their decisions on broad national strategy. Research and development resources are much more narrowly distributed in Britain than in most other countries, with an abnormally high concentration on military uses and nuclear power. The UK government spends more than half of the total national R & D expenditure, 70% of which is devoted to defence and nuclear power with less than 10% of taxpayers' money being applied to other forms of civil wealth-creating activities. In his book *Innovation and Employment* David Foster estimates that although 80% of national wealth creation in Britain appears as spending power for individuals, only 2% of innovation relates to new products for individual purchase: 72% is directed towards capital goods. These figures illustrate the distortion caused by Government and show how technology has become the creature of business rather than the servant of people.

This concentration of technical resources in big firms and big institutions would perhaps be somewhat less objectionable if it provided the maximum amount of innovation for every pound spent. Regrettably the reverse is the case. An investigation by the US National Science Foundation showed that small organizations were four times as productive as medium-sized ones and twenty-four times as productive as big ones. In Britain a double penalty is being paid. Not only are wide areas of technology neglected because of the gross maldistribution of technical resources, but a great disadvantage is suffered because most of the resources are concentrated in the kinds of big organization which are the least productive of innovation. The consequences of the way in which technology choices are made in Britain are to some extent illustrated by the trends in import penetration shown in the table opposite.

Technologies for people

If a new and more appropriate direction for technological development is to be evolved – providing prosperity to individuals and communities with much less adverse cost and being people orientated rather than based solely on business profitability – it is necessary first to sketch the kind of possibilities that exist. Before doing so we need to be reminded of the criteria and constraints which apply to a technology 'as though people mattered most':

1. Life enhancing in both the short and long term; tailored to meet the full range and variety of human need – material, intellectual, emotional, and spiritual.
2. A scale as small as possible consistent with the efficient use of material, energy, and human resources.
3. Economizing in the use of resources – materials, energy, land, plant and animal species – at levels of consumption that are sustainable in the long term for the growing world population.
4. A minimum level of capital intensity for a complete system of production and distribution consistent with the efficient use of all resources.

Item	Import penetration 1976 (%)	Increase in imports 1968-76 (%)
Pharmaceutical	20	100
Synthetic materials	35	60
Agricultural machinery (except tractors)	45	87
Textile machinery	74	94
Earth-moving equipment	61	97
Mechanical handling	23	92
Office machinery	95	29
Document-copy equipment	100	118
Watches and clocks	74	23
Surgical instruments, etc.	45	136
Scientific instruments	41	64
Electrical machinery	20	150
Telephone apparatus	14	250
Electronic components	50	67
Computers	69	50
Domestic electrical appliances	27	146
Motor-vehicles (cars 1968-78)	45	800
Motor-cycles, cycles	74	90
Textiles (all groups)	28	75
Leather goods (not footwear)	33	106
Outer clothes	26	270
Underwear	41	95
Footwear	24	72
Plastic finished products	17	70

Source: *Innovation and Employment*, p. 116

Because an 'appropriate technology' is concerned with satisfying human need it will pay regard to the main categories of expenditure in an average British family:

Shelter – housing, fuel, and electricity	20%
Food and drink	30%
Clothing and footwear	10%
Durable household goods	10%
Transport and vehicles	15%
	85%

leaving 15% (or slightly more) to be spent on family interests, recreation, holidays, newspapers, television licence, presents, and so on. (It must be remembered that equally important items such as education, roads and sewers, public amenities, health care, defence, social security, and pensions are provided out of all forms of taxation which in aggregate is equal to about 44% of gross national domestic product. Appropriate technologies must operate at a level of efficiency that will provide sufficient surpluses to cover these costs also.)

Shelter

Under the existing economic system there has been a marked increase in house costs. In 1945 a three-bedroom semidetached house could be purchased by a young first time buyer with the equivalent of two years' earnings. At that time mortgage interest rates were about 5%. In 1983 a similar property will cost three to four years' earnings and interest on mortgages runs typically at about 10%: it becomes increasingly difficult for first time buyers to meet these costs on a single income. Although running costs have not grown to such an extent, largely because for most of the post-war period energy became cheaper in real terms, they are now increasing more rapidly as the real cost of gas, coal, oil, and electricity goes up. Piecemeal efforts have been made to reduce some of the costs of both the house and its running costs, but no concerted effort has been made to tackle the overall cost in a radical way.

Dr Freddy Clark of Harwell has produced a useful way of illustrating the interrelationship between capital and energy costs in home heating with various different heating systems (see Figure 6). The line AC and those parallel to it represent lines of constant total annual cost, which increases the greater the distance from point B.

The x marked 2 on the figure represents a poorly insulated small house with gas boiler heating. The x marked 1 represents the same house which has been well insulated. Although there has been a small increase in the annual capital charge to cover the additional cost of retrofitted insulation, the reduction in energy cost is so much greater that there is a reduction in the total annual cost. Points 4 and 3 represent the same house, without and with insulation, with gas at double the existing cost in real terms. Under these conditions, which are likely to apply by the end of the century, the reduction in total annual cost resulting from insulation is greatly enhanced.

Other methods of achieving the desired internal temperatures throughout the year can be plotted on the same chart. Different house construction methods, which affect the thermal properties, can be

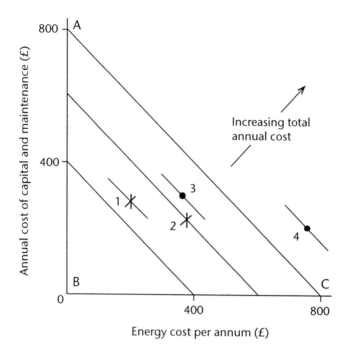

Figure 6 Relationship between capital and energy costs in home heating for various systems

included, as well as different energy sources, including natural sources such as wind, solar, geothermal. From such a plot the benefits of building houses with improved thermal properties in the face of escalating real costs of energy become evident.

For centuries buildings were massively constructed. This provided a thermal flywheel effect which helped to maintain a fairly constant internal temperature despite changes occurring out of doors. After all, one of the basic purposes of a building is to provide a comfortable and constant indoor climate. With the appearance of plentiful supplies of cheap coal, structures became much less massive. Insulation and the thermal flywheel effect were lost and it became necessary to use increasing amounts of fuel to maintain the required indoor climate; initially in one room only, with the remainder becoming distinctly uncomfortable in winter. When central heating was introduced into these thermally inferior structures little was done to improve the standard of insulation by retrofitting. As a result there now exists in Britain a stock of buildings, poorly designed from a thermal point of view, which are increasingly expensive to heat as energy prices escalate. A new, low-cost form of construction is required which has a very high level of insulation, controlled ventilation, is massively constructed to provide a thermal flywheel effect, and is designed to maximize the solar heat gain. 'Passive solar design', in contrast to the solar panel collectors used for water and space heating, is of the utmost importance.

The development work done by John Parry, a building systems and materials consultant and Chairman of the Intermediate Technology Development Group Building Panel, provides the basis for such a new form of construction. Initially developed for tropical Third World countries, it has been applied to some experimental single-storey buildings in Devon, where the 'cob' form of construction was used for several hundred years. Cob walls were built of layers of shaly local clay mixed with straw and animal dung. They were protected from the weather by coats of limewash and a generous roof overhang to shed water clear of the face of the wall. John Parry's invention is based on bricks made from local clay which are dried but not fired – saving both transport and energy. These unfired bricks are laid unmortared on a conventional foundation and fired bricks up to the level of the damp proof course. A layer of glass fibre reinforced cement (grc) is laid on the damp proof strip and, as the wall is built, it is rendered with a continuous layer of the same grc. This rendering, which encases the whole wall, provides structural bonding to the bricks, gives rapid protection of a high order, and is visually attractive. The following comparisons show the very considerable benefits of financial cost and energy cost which can be obtained with this form of construction in Britain.

75 sq. metres of 13" wall	Financial cost (£)	Energy cost (kw)
Conventional brick wall	731	43,176
Unfired clay brick wall	307	3,660
4" concrete wall	487	723
Cement stabilized soil wall	240	6,070

Experience in Devon with unfired clay bricks suggested that cement or lime stabilized soil would on several counts be preferred in British conditions. With an energy cost of about one-seventh of a conventional fired brick wall and a financial cost of about one-third, there is obvious scope with this construction method to build massively thick walls, gaining insulation with no increase in cost.

Another potential benefit of this method of construction is obtainable when linked with a passive solar concept pioneered by a Frenchman named Trombe. By locating a glass outer skin on the south facing wall of a building and circulating air from the house through the gap between glass and wall, heat which has been retained in the wall as a result of solar radiation and the 'greenhouse effect' is transferred to the circulating air and is used to warm the house. In the Wirral, in the cold and overcast north-west of England, experimental houses which incorporate this concept show a halving of fuel consumption compared with identical houses that do not have a Trombe south wall. Since with this concept the south wall is used as a solar energy store, the value of a massive construction such as the one described above can be easily appreciated. A new type of glass, produced by Pilkington Brothers of St Helens under the trade name KAPPAFLOAT, produces an enhanced 'greenhouse effect', giving a greater solar thermal gain through windows as well as in a Trombe wall application.

It is well established that a high standard of insulation and controlled ventilation (with or without heat exchange) can more than halve energy consumption. When a Trombe wall is used a similar saving is made. A further halving can be obtained by substituting a heat pump for a conventional heating system. By adopting all these measures it is possible to envisage a six- to eightfold reduction in energy consumption for space heating, which accounts for about 80% of a household's energy bill. By using the alternative construction system this saving should be obtained at very little extra capital cost of the building and its heating system. A saving of between £250 and £300 pa on the total cost of, say, £1,800 pa for mortgage, energy rates, and maintenance is one that would be a great benefit to an average household, and would be even more appreciated by those people being impoverished by escalating energy prices. The saving would, of course, increase as the real price of energy increases.

Another feature of the John Parry building method is rapidity of construction and modest skill requirement. It therefore lends itself to 'self-build' construction schemes. In 1981 self-builders erected 7,000 houses. This approach to the acquisition of a house can result in a saving of between one-third and one-half. Consequently a much smaller mortgage loan is associated with self-building. Thus the combination of energy cost saving described above and the capital cost saving of self-build would provide housing at overall annual outgoings much lower than with the traditional and wasteful system. As the amount of time that people have to spend in paid employment shortens, more people are likely to be attracted to building their own houses in their spare time. A sabbatical year or half-year, perhaps taken in the mid-twenties, would provide a most valuable opportunity for such a project.

Food and drink

Food and drink still constitute the biggest item of expenditure for the average British family. Farming and the food and drink processing industries have undergone a revolution in this century, and still the cost absorbs nearly 30% of the increased family average earnings. Why is it still such an expensive item on the household budget?

Direct comparisons of household expenditure on food and drink between 1900 and 1980 are meaningless, because in 1900 an average working family consisted of six individuals whereas in 1980 there were only three. On a per capita basis expenditure was £0.19 in 1900, in 1980 it was about £9.00 (i.e., 45 times as much as in 1900, compared with an inflation factor of about 35). This modest increase in real terms is in part accounted for by an improvement in the quality of diet. A comparison of prices of individual items suggests that the increased cost a head is also to some extent a result of real price increases. In 1980 the price of bread per pound was in the range of 25-28 pence compared with six pence in 1900. Applying a rate of inflation of 35 to the 1900 price would put it at 21p in 1980 money. Despite the vast changes that have taken place in farming, milling, and baking, involving vast capital, material, and energy expenditure, the real price of bread is not less than it was at the beginning of the century.

There is little argument that better-quality bread can be bought from small local bakeries, or baked at home, but there is a valid objection that it is more expensive than the mass-produced product. The major bread producers also supply most of Britain's flour and they charge small bakeries (and the domestic purchaser) an inflated price for flour to offset the big discounts demanded by supermarkets on their orders for factory-made bread. Small bakeries are in effect subsidizing the major bread companies. With efficient small mills operated by farmers – thereby increasing their added value and reducing their vulnerability to the squeeze on prices applied by the big companies – flour could be supplied directly to local bakers and retail shops at fair prices.

Elimination of the big price differentials that favour big companies at the expense of smaller ones is of fundamental importance to the development of a more appropriate production and distribution system. Its achievement will require legislation to outlaw discounts and rebates greater than can legitimately be justified by genuine economies of scale in distribution. Such legislation would ensure that householders obtained fairly priced flour regardless of whether it came from a farm which milled its wheat or one of the big producers. The same principle applied to all basic foods would change the price relationship between industrially processed food and drink and home or small scale processing. Because of the unfair differentials that currently exist in the prices paid by house-holders and industrial processors, it sometimes appears that there is little difference in retail shop prices between basic fresh foods and the industrially processed product. The high costs of production, packaging, and distribution of the industrial product are largely paid for from the lower price that the industrial processor pays for his raw materials. When these differentials in basic foods are greatly reduced, householders will see a much greater cost advantage in home processing, and small local food processing concerns will compete much more successfully with the big national companies.

During this century the agricultural industry in Britain has made strides in reducing the cost of most farm products. As food processing has increasingly been industrialized – twice as many people are now employed in the food and drink processing industries as on the farms – the lower costs of basic foodstuffs have largely been denied to householders. When those lower prices become available a substantial reduction in the average family food and drink bill will be enjoyed by those families that choose to do some of their own food processing.

The magnitude of saving potential will vary from one product to another as we move away from the present highly industrialized system of processing and distribution. Milk provides an interesting example. In 1984 it was collected from farms around Corby Glen in Lincolnshire and taken by bulk tanker to Peterborough for processing and packaging. The milk delivered to householders in Corby Glen is a different batch, coming from Nottingham. When it leaves the farm gate it is at a price of 8p per pint – paid to the farmer. When it arrives on the doorstep it costs householders 20p per pint. Some of the difference is accounted for by the high overhead costs of a highly industrialized and centralized system, but a great deal results from the cost of packaging and transport.

In some food and drink products the cost of packaging is not much less than the intrinsic cost of the materials inside. Indeed in some cases, such as bottled soft drinks or beer, there is such a high content of water in the product that the container represents the highest single material cost element in the packaged product. Worthwhile cost savings can be made with distribution systems that minimize packaging costs. Moreover, if the world's stock of timber is to be saved there will have to be a big reduction in the use of paper products in packaging.

Energy consumption in processing and distribution provides an interesting illustration of how one cost element, which is present in all stages, adds to the final price of a product. In his book *Fuel's Paradise* Peter Chapman estimates that of the 17.5% of primary energy input to the UK accounted for by the food industry, 13.1% is taken up in

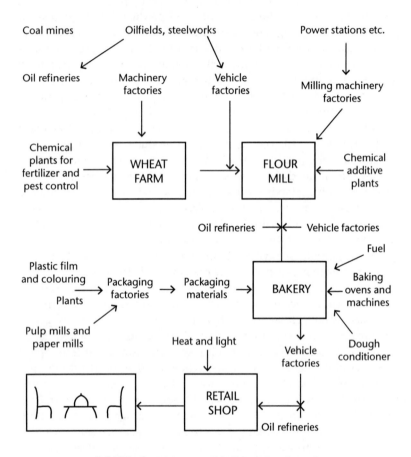

5.6 kWh of energy are used in this system to produce and distribute ONE 2lb. LOAF OF BREAD

Of this TOTAL 5.6 kWh the FARM uses 20%
the MILL uses 15%
BAKING uses 23%
TRANSPORT & PACKAGING uses 22%
RETAILING uses 20%

Figure 7 The life history of a loaf of bread

activities that are ancillary to food production and processing itself. In other words, three times as much energy is consumed in packaging, transporting, and distributing food as is used directly in its production and processing.

It is not, of course, only energy that can be saved by a simplified system compared with the one illustrated in Figure 7. Several million tonnes of domestic waste, a considerable proportion of which derives from food and drink packaging, must be disposed of each year in Britain. In her book *Material Gains* Christine Thomas notes that over a 40-year period the quantity of domestic waste increased by 70%. Whereas in past times the garbage was mainly organic (plant and animal) waste, much of the increase that has occurred is tinplate, glass, paper, and plastic. A household which either purchases in a shop preserves (such as sauces, jams, and marmalade) from open bulk supplies which are dispensed into private containers, or else carries out preserving in its own kitchen, can largely eliminate glass from its domestic waste. Similarly, by carrying out more food and drink processing of all kinds at home, much of the other packaging waste can be avoided. As paid working hours shorten more time will be available for the varied and satisfying work of food and drink preparation and preservation, and there is no lack of suitable domestic kitchen appliances to ease this type of work.

Even if a great deal of this type of work returned to the domestic kitchen, there would still remain a considerable demand for commercially processed products. Although there has in recent years been some attention given to small bakery plant development, there is little modern commercial scale equipment available for most other local small-scale food and drink processing. When, for example, small new local breweries have opened they have often been forced to operate with obsolete equipment from abandoned plant.

The present highly centralized large scale food and drink processing system has given rise to a mammoth increase in road transportation and its associated environmental disadvantages, including more injuries and deaths in road accidents. The Annual Abstract of Statistics reveals that there has been a steady increase in the volume of freight carried since the Second World War and a similar steady increase in the distance that each tonne of freight is carried. The decentralization of commercial food and drink processing, combined with a return of a proportion of such work to domestic kitchens, could reverse both these trends in transportation statistics, saving both costs and lives. The absurdity represented by a truck-load of biscuits travelling south on the Ml motorway being passed by a similar load going north must disappear if sanity in economic affairs is to return.

The system of mass distribution and retailing through supermarkets has developed in parallel with the highly centralized production system. Its ability to replace local, small specialist food and drink shops has depended primarily on two things.

Firstly, the big price differentials made possible by large discounts favour the big retail chains in comparison with small local shopkeepers. Secondly, supermarkets reduce their shop floor labour costs by making their customers carry out unpaid part of the work which would otherwise be done by employees. The main attraction of the supermarket, however, is not so much lower prices – frequently the difference in the cost of a basket of products is marginal – but the convenience of shopping for all products in one location, particularly when there is a customers' car park. The revival of small specialist retailers may well require the emergence of a modernized version of the Victorian market hall in which there is independent ownership and stocking of product shelves with customer self-help in shopping and a single combined payment at the exit. With the availability of modern electronic recording systems 'stallholders' could be credited daily with the proceeds of their sales as registered at the check-out. Such a system would provide easy access for small scale local processing companies to a modern system of distribution and retailing such as is impossible for them in the present system of retail supermarket chains.

In the change of direction here envisaged for food and drink, agriculture would not survive in its present form. The development of large scale processing and distribution has had a major influence on farming, and a decentralization to smaller scale local processing and distribution would also make important changes. In the same way that the processing and distribution systems were the product of an era of very cheap energy, modern agriculture has become heavily dependent upon fossil energy for the operation of machines and plant and for the manufacture of oil-derived fertilizers and biocides. If food and drink prices are not to increase with the real price of energy, alternative methods of agriculture less dependent on energy will need to be introduced. There are two very good examples of such alternatives which have proved successful over a period of years. The late Sam Mayall's farm near Shrewsbury has been farmed with very little use of chemicals and with above average crop yields for some thirty years, and profitability has been enhanced by on-farm milling of grain for distribution direct to small retail outlets. Neil Wares's Bore Place Farm near Edenbridge in Kent is completely energy self-sufficient by means of a very large methane digester which converts cow dung to gas and an excellent fertilizer. His 156 hectares of farmland provide grazing for 350 Friesian cows and produce some 2,800 tonnes of silage each year at a gross profit of £468 per hectare. Savings on energy expenditure produced by the biogas digester have run at nearly £3,000 per annum and are expected to increase further to £4-5,000 with further improvements. The residual manure slurry obtained from the digester is estimated to save £5,600 pa in commercial fertilizer. The combined savings on energy and fertilizer make a very worthwhile contribution to the total gross profit of £73,000 at present prices; the proportion will be even more significant when the real price of energy increases.

Realistic alternatives to the way that food and drink are made available can, in an era of shorter paid working hours, not only provide a sufficient and varied supply of products at reduced cost but can also make worthwhile contributions to resource economy and environmental improvement as well as saving lives and injury from road accidents. All of this in the biggest cost item in the average family weekly budget.

Clothing and footwear

Although clothing and footwear accounts for only about one-third of the expenditure on food and drink it is felt by most people to be a very important sector of a modern family budget, giving a great deal of pleasure as well as performing practical functions. The past half-century has seen extraordinary advances in the technology both of textile and footwear materials and of manufacturing processes for materials, garments, and shoes. Modern textile machinery represents some of the most remarkable achievements in mechanical engineering.

Looking to the future it seems very improbable, even with much higher costs of oil and other fossils fuels, that synthetic fibres will become too expensive for widespread use. In the long term, feedstocks to synthetic fibre plants may change from oil to coal and ultimately to biogas (methane), but it seems unlikely that the fibre products will be manufactured in plants which are very much smaller than present ones. With the modern processes of dyeing, spinning, knitting, and weaving, with both natural and synthetic fibres, there is no longer any reason for industrial plant to be concentrated as the traditional woollen and cotton industries were: these processes are now capable of being carried out efficiently on a comparatively small scale. Some experience in recent years shows that in fashion influenced businesses there are real advantages in avoiding the rigidities of big scale centralized units.

In parallel with the decentralization of all the processes apart from synthetic fibre production there is an opportunity for a considerable revival of a modernized form of craft spinning and weaving and garment production, both as a domestic (DIY) activity and also as a form of small business carried out either full-time, or part-time in conjunction with some other form of income earning. Developments which have been pioneered by the Intermediate Technology Development Group, aimed at markedly increasing the productivity of village spinners and weavers in India and Bangladesh, point the way to further machine development for the benefit of craft workers in Britain. With the excellent domestic knitting and sewing machines that are already widely used, developments in spinning and weaving machines should make possible a considerable revival of craft products and garments. With the use of microelectronics goods of high quality can be produced by a comparatively unskilled person. To enable this change in the provision of clothing to reach its full potential it will be necessary for basic textile fibres and yarn to be available at prices which differ very little from prices paid for the same products by industrial producers.

It is frequently alleged that textile and garment industries cannot compete in a high-wage economy with industries in low-wage countries. This argument is based on the mistaken view that in a modern production plant direct shop-floor wages represent a high proportion of the total product manufacturing cost: in fact it usually represents between 5% and 15%. Providing that the labour costs of overhead activities are kept well under control – as is usual in small firms but not in large ones – the difference in manufacturing cost between high and low wage levels is small, generally smaller that the high costs of transportation and warehousing that are necessary in bringing goods half-way round the world to the British market.

In textiles a large proportion of the cost of finished products is in the materials used, and the cost of direct, manufacturing man-hours is very small: the small difference in retail price between the wool required for home knitting and the price of a commercially knitted pullover illustrates the point to some extent. At present energy price levels (oil feedstock prices for synthetics) there is no intrinsic cost disadvantage to prevent commercially competitive textile and footwear production in Britain. Indeed it is carried on in countries with wage levels considerably higher than they are in Britain – in the USA, Canada, Sweden, Switzerland, for example. The competitive position will become even more favourable as transport costs reflect the increasing real cost of oil.

Such changes in the ways that clothing and footwear can be provided in the future do not promise a significant reduction in cost, but they should provide a hedge against an increase in cost which will occur if no change is made.

Durable household goods

The purchase of house furnishings and hardware, such things as cookers, washing machines, cleaning machines, garden mowers and tools, refrigerators, freezers, telephones, sewing machines, television, radio, and hi-fi systems, to which must now be added video sets and home computers, amounts to 10% of average household expenses. At any one time the total replacement value to one household of all this equipment would be in the region of £5,000; it would be a lavishly equipped average home that contained items whose total value exceeded £10,000. If total annual average family expenditure is of the order of £5,000 then a 10% rate spent on durable household goods implies an average life of 10 years only for contents valued at £5,000. This is not surprising so far as domestic machines are concerned. Many break down within ten years and are not repaired; but it is surprising that furnishings and hardware are being replaced with the same frequency, particularly as most of it has a very much greater life potential. Well made tables, chairs, and cupboards have almost unlimited durability.

The choice that is faced in this category of expenditure is whether we continue to make and use household goods of poor durability – products of a mass volume and minimum first cost system – or whether we adopt the alternative of more durable goods, which although they may cost slightly more initially will have a substantially lower lifetime cost per annum. Doubling average useful life from 10 to 20 years would not greatly increase first cost, but it would nearly halve outgoings on household goods.

There are signs of a change beginning in the furniture sector with a revival of craft made items. Where such products come from top designers they are priced outside the range of an average family; but most of the goods that are being produced by less celebrated craftsmen are sold at very reasonable prices.

There are no parallel signs of more durable domestic machines being produced. That objective can probably best be achieved if high standards of durability and repairability are built into the British Standard specifications for such machines and sales are restricted to models that comply with these standards.

With the availability of a greater amount of leisure time there will be a splendid opportunity for people to enjoy making some of their own furniture. Also with proper design attention, and with the provision of carefully prepared information, it should be possible for much routine servicing and minor repair work to be carried out by householders on domestic equipment. This could make a valuable contribution to the application of the 4Rs to scarce resource conservation.

Transport and vehicles

At the beginning of the century travelling was virtually a non-existent item in the average family budget. Now it accounts for almost as high a proportion as shelter (15% compared with 17%). Widespread car ownership is the principal cause of this large item of expense; and it would be even higher were it not for the fact that most average families are running secondhand cars in which a substantial amount of the high initial depreciation in value has been paid for by the initial owners, often companies.

The following is a rough breakdown of £750 (15% of a total annual expenditure of £5,000) for a 4-year-old car costing about £2,500.

Annual road licence	£85
Car insurance	50
Depreciation (over 10 years)	250
Servicing/repairs	95
Petrol and oil (over 5,000 miles)	270
	£750

If a legitimate charge for garaging is made the cost increases to about £850 pa or 17p per mile. Surveys have shown that a substantial proportion of annual mileage is on very short journeys (3 miles or less) and for only a comparatively small proportion of mileage is there more than one person in the car. The low annual mileage indicates that this very costly family investment is only in use for about 200 hours each year, or 2000 hours over the lifetime of an average car.

Clearly mass car ownership is a very costly and inefficient way of providing people with transport: it can, too, be blamed for most of the injuries and deaths caused by transport accidents. To a considerable extent it is private owners that use their cars for very short journeys, many of them around town where most accidents occur, where petrol consumption is highest, and where parking is a considerable public nuisance. A combination of alternative transport for short journeys (bicycles/tricycles – with or without power assistance, frequent taxi-type small bus services such as have become very popular in Belfast) and car hire for longer journeys with several occupants, would be much more rational, no less convenient, and less expensive than the present system. As well as the great benefits of fewer accidents, less congestion, and reduced pollution in built-up areas, there would be the added advantage of a transport system common to the entire population, eliminating the class distinction between car owners and non-owners. The increase in cycling, the decreasing average mileage of cars, and plateauing of car ownership all point to the beginning of a change which is likely to develop as an alternative to car ownership becomes more attractive.

Between 1966 and 1974 the average life of a car in Britain decreased from 15 years to 10.7 years. However, since 1974 the trend has reversed, so that in 1983 average car life is about 13 years. This shift from manufacturing to servicing, repair, and reconditioning is desirable because it provides better quality employment in small dispersed units as well as reducing the consumption and waste of materials and energy. A move from private ownership to car hire would further increase servicing at the expense of manufacture. Although average car life in years would not increase, mileage covered during a car's lifetime would because the need for hire companies to gain the maximum benefit from their investment would lead to more intensive use of each vehicle.

CHAPTER 5

Engineering the future

In Britain the term 'engineer' is widely misunderstood and misused. Most commonly it refers to all categories of manual workers in manufacturing industry, as though all the people who work in hospitals were called surgeons. The true meaning of the term, as it is used worldwide, refers to those who 'apply their genius', or 'work with ingenuity', in designing things which harness the resources of nature for the benefit of man and for the good of the environment. In many European countries 'Engineer' is a title of great honour and gives high status to the profession in these countries, where recognition is given to the central part that engineering plays in the developments of modern civilization. While due honour is paid to the knowledge and skills of those who practise modern surgery, it is not well understood that without the special medical engineering equipment that is the product of the ingenuity of engineers surgeons would be helpless. If economic development is to change direction in the way that is argued in this book, engineering will play a central part in the process. The genius – the ingenuity – of engineers will need to be redirected towards different objectives.

Nature's engineering

Ingenuity can be applied in many different ways, as is evident from the processes of nature. The superb photography of many television nature programmes provides a brilliant demonstration of the extraordinary variety of means used by plants and animals to sustain themselves in their particular local environment. In thinking about the future of engineering it can help to contrast some of the characteristics of engineering as it has been practised in the past with nature's engineering.

A most striking characteristic of nature's engineering is its subtlety and sophistication. From a comparatively small number of chemical elements it evolves a myriad finished products combining atoms into molecules; molecules into crystals and cells; crystals to rocks, and cells into fibres and muscle and bone in a fully automated production system. It is a system that is in continuous operation, working steadily by day – powerfully but noiselessly – and at night it rests. Until its design life is completed it does not break down; and such faults as develop from time to time are self-corrected by inbuilt mechanisms.

So long as the sun shines it needs nothing else to drive it; and it never wears out because it constantly renews itself, working in harmony as an integrated system and recycling all of its waste products. At all times this amazingly complex and interactive system is in equilibrium and yet each part is in a perpetual state of development. Despite the fact that the production system is fully automated, each individual example of every production is unique.

There are other significant observations to be made about nature's engineering. It operates within a narrow band of temperatures and pressures with maxima only a little above ambient. Consequently the processes operate at a slow rate, even when some form of inbuilt energy storage permits a brief high-speed burst of activity such as the flash of lizard's tongue as it captures its prey. A further important point is that the geographically dispersed nature of the system's operation avoids the concentration and accumulation of quantities of waste products which 'natural' recycling processes are incapable of handling. Finally, it is noticeable that there is an avoidance of uniformity of scale in nature's engineering products and processes.

Characteristics of modern engineering development

If we follow the ways in which engineering has developed over the past two hundred years of the industrial era we see that it contrasts very sharply with nature's engineering, growing further away from it as time passes with increasingly damaging effects.

The key factor in the development of engineering is the use of nature's limited store of highly concentrated energy locked up in fossil fuels – coal, oil, and natural gas – in place of diffuse direct solar energy. By burning these fuels and creating very high temperatures and pressures, engineers have been enabled to produce materials technologies and plant and machine designs which have resulted in bigger, faster, and more powerful machinery and processes. A walk around the Science Museum in London or the Railway Museum in York illustrates very clearly the progress made in power, speed, and size. What is not so apparent from such an inspection is the effect of these trends on people and the environment. The pressures on people in their working lives have become so great that some continuous work is limited to very short periods and many people are unable to maintain the required pace of work at a comparatively early age. The development of industrial production into bigger units has contributed to the concentration of population into a few big metropolitan areas with all the noise and other environmental problems adding to the psychological effects of a very dense and congested habitat. Environmental damage, whether it be noise, air, water, or land contamination, is directly related to the concentration of waste product emissions. Nature is capable of dealing with pollutants in small quantities widely spread: its remedial processes are overwhelmed when huge quantities are emitted over a small area.

In the early years of the industrial revolution manufacturing plant and machinery required considerable skill to operate. In a sense they were improved tools which increased an operative's productivity and reduced physical effort, but retained and in many cases heightened skill requirements. With the passage of time, however, technological development has increasingly de-skilled industrial work, making the human qualities of workers largely redundant, to such an extent that there is disharmony and alienation between man and machine.

For the first 150 years of the industrial era products were still designed for long life and ease of maintenance and repair. But as the pace of production increased the concept of 'produce and throw away' took over, a notion that was greatly encouraged during two decades in which both energy and materials were cheap.

Another important trend that can be observed in retrospect is one of increasing complexity. In some industries, such as aviation, the costs now involved in the development and production of a major project, an airliner or even an engine, can no longer be borne by a single European national industry and an international consortium will have to be formed to undertake it. Early motor-cars were developed and produced in a local garage: the investment in a new model of the complex modern successor is several hundred million pounds, a level of investment that must be justified by a world market. In a world that becomes increasingly nationalistic this is surely the road to absurdity. That absurdity is perhaps best illustrated in relation to commonplace traditional things. Simple arithmetical computations were carried out either by rapid mental processes or with the assistance of an abacus. Now a highly complex industry produces electronic calculators to perform these simple functions, making mental arithmetic redundant and shop assistants helpless when their machine breaks down.

Perhaps the most surprising thing about engineering development is how little progress has been made in improving the efficiency of resource utilization, with the sole exception of the human input. Whereas human productivity has been increased by a factor of more than ten, we still only manage to achieve efficiencies in our use of fuels of about 20% in road transport and 35% in electricity generation. The waste of metals and other raw materials, furthermore, increases at an exponential rate. The technological system that we have engineered displays the ultimate absurdity of making redundant human resources that are in plentiful supply, while it indulges in the profligate waste of material and fossil energy resources that are in limited supply. By departing so far from the characteristic pattern of 'nature's engineering' we have created a system that is not sustainable and which, even though it is restricted to about a quarter of the world population, is steadily laying waste much of the global natural environment.

In his book *The Clever Moron* (1977) R.S. Scorer wrote:

> We are heading for a period of human misery on a scale unprecedented. Throughout the long history of humanity until the industrial revolution, it seems that wealth was accumulated very slowly, and the rich were those who had plundered by conquest. Then Industrial Revolution Man seemed to change all that; for wealth was created without taking it from another man, or so it seemed. But in reality we were stealing it from posterity. The accounting is about to begin. Man has been prodigal because he has been clever, but his indulgence has made him a Moron.

The questions that this generation has to ask and answer are: 'Do we have to continue like Gadarene swine to engineer the future in this crazy way? Are there no different ways in which we can use our ingenuity (engineering) that will be sustainable and provide the global population with a chance of a full and satisfying life?' In the following paragraphs some examples are given which help to illustrate the directions in which to search for solutions. Of themselves they will not provide a 'technical fix' to the problem, but they can provide the alternative means which give scope for radical social and economic change.

Resource economy and environmental well-being

A major slowing down in the rates of material and fossil energy consumption in the industrialized world, and a dispersal of activities and population, are necessary conditions for improving conditions in the Third World and at the same time avoiding catastrophic environmental damage.

For engineers this means, first and foremost, design and development with much greater emphasis on durability, repairability, ability to incorporate important new developments into existing equipment, and ease of disassembly and recycling of the materials of construction. But engineers will only be able to adopt such new design briefs if new elements are incorporated into the national and international standards systems which must be satisfied if a new design is to be acceptable. A crucial responsibility rests upon the official bodies which set standards, and engineers play an important part in the work of such bodies. They frequently do so, however, as employees of industrial concerns, whose interests may not be best served in the short term by such changes as are here envisaged. It is desirable, therefore, that engineering should be represented on standardization bodies by independent professional engineers who have the support of their professional institutions.

Although in recent years the application of modern technology to repair, reconditioning, and recycling has received increased attention from engineers, these activities are still at a comparatively primitive stage of development. However, in contrast to production processes they are likely to remain labour intensive, requiring a considerable amount of skill and dexterity. Attention will need to be paid, therefore, to establishing and maintaining much higher standards of skill and competence than are currently present in this neglected part of the economy. Just as engineering technology has enhanced the doctor's diagnostic skills, development in machinery fault diagnosis could be a valuable aid to the engineering technician.

The trend in recent years whereby machine repairs could only be effected by the replacement of complete expensive units has placed repair at a considerable economic disadvantage. A new approach to component design that will allow the easy replacement of individual defective parts will need to be one of the requirements of standardization specifications.

It is often argued that an extension of the useful life of a particular manufactured item may be counter-productive because it will delay the introduction of new developments which would be more economical of resources. Taken to its logical conclusion, this argument should mean that every article should be scrapped each time a new modification becomes available. There is, however, a point in the contention which needs to be recognized by the designer. A good design – well illustrated by the Spitfire aircraft which over a period of some eight years of continuous development was transformed in its performance – can provide ample scope for incorporation of foreseeable technical developments. High durability and long service life need not become barriers to development.

In striving for minimum first cost as an overriding priority, intrinsic product durability has frequently suffered. There is no lack of knowledge as to how durability could be improved, but a price must invariably be paid; and engineering as it is applied to commercial affairs is about how a technical objective can be attained at minimum additional cost. If intrinsic product durability is to be given a high priority in engineering design, a great deal of work will need to be done to develop optimum solutions in each application.

Although the increase of emphasis on repair, reconditioning, re-use, and recycling at the expense of production of goods will greatly facilitate the dispersal of activities and population and reduce environmental damage, attention will also need to be given to production engineering developments which permit product manufacture to be carried out in smaller dispersed units. The processes which are least amenable to such a development are those used for the production of basic materials – metals, cement, chemicals, glass. However, this does not mean that these industries and their engineering will be unaffected by a change to a conserver economy. A consideration of likely effects on the steel industry will serve as an illustration.

The British Steel Corporation's Ten Year Development Strategy agreed in 1973 forecast an annual steelmaking capacity of 36-38 million tonnes for the early 1980s. Since then demand has fallen to 15m tonnes and is likely to fall even further even without a fundamental change in economic development. In a conserver type of economy the fall in demand over a period of about thirty years would be very dramatic, to something like 5m tonnes per annum. The change, however, would not be merely quantitative. There would need to be a major change of feedstock material, with a considerable proportion of iron ore, and iron making, being replaced by scrap. This inevitably will require design and development of steelmaking plant to accommodate such feedstock. The technical quality requirements of the steelmaking processes will also markedly change as a result of changes in the requirements of steel users. In an article in the *Lloyds Bank Review*, Jonathan Aylen, an economist specializing in the steel industry, wrote:

> Perhaps the worst legacy of the Finniston years (in BSC) was the belief that steel is a standard product that does not have to be marketed. Briefly, those within the Corporation who advocated development of high value added, sophisticated products lost the argument to those who advocated volume production in order to realize growth and economies of scale.

Neither growth in output nor economies of scale have been realized in a declining market, and opportunities to meet demand for specialist steels (e.g. for North Sea pipelines) are being lost to foreign competitors. British Steel starts from a position in which existing plant does not make a good match with technical quality demands; and changes in demand which will result from future 'conserver economy' developments are likely to transform BSC from a high volume, low added value, bulk steel producer into a much more complex producer of high quality special steels: demand for greater product durability, for example, will require more corrosion-resistant alloys and various types of coated steel. In the important automobile industry the search for weight reduction in vehicles will not only displace considerable quantities of low grade steel with synthetic plastics, resins, rubbers, etc., but such steels as remain in use will be lighter weight sections with higher strength. The Mercedes 90 is a first step in this direction. It obtains good impact resistance – in case of accident – from a lighter body made from thin, high strength alloy steel which also saves fuel and enhances performance. The steel industry is very energy intensive. Consequently as the real price of energy increases, as inevitably it will as traditional cheap sources become scarce, there will be an even greater pressure on steel plant engineers to develop processes which are much more thermally efficient. Although the impact of a change to a more conserving form of economy will affect each material producing industry differently, all will similarly experience profound effects.

In the production engineering of parts, components, and finished products the trend is likely to continue away from fully integrated plant, with raw material input and finished product output, to geographically separated and independently managed operations. This change should not of itself create a need for a radical re-engineering of production processes – present integrated plants are agglomerations of independent but inter-related production units and machines. The need for a new engineering approach is in the interests of creating employment for production workers which is more fit for human beings. In a paper published in 1981 in the Institute of Electrical and Electronic Engineers Control Systems magazine, Professor H.H. Rosenbrock identified a basic difference in the attitude of production plant design engineers towards machines and equipment on the one hand and employees on the other. He pointed out that while engineers go to great pains to see that there is as near perfect as possible a match between a machine's capacity and the function it is required to perform, no such consideration is given to the employee's potential capabilities and the job that he or she performs. As a consequence there is very frequently in modern production engineering a gross mismatch between job and worker involving a massive under-utilization of human ability and adaptability. He quotes in his paper an extract from Frederick Herzberg's book *Work and the Nature of Man*:

> The tasks assigned to workers were limited and sterile ... the worker was made to operate in an adult's body on a job that required the

mentality and motivation of a child. Argyris demonstrated this by bringing in mentally retarded patients to do an extremely routine job in a factory setting. He was rewarded by the patients' increasing production by 400 per cent.

The indifference of production plant designers towards the job/worker match not only – and most importantly -- dehumanizes workers with normal capacities, but also ultimately impairs plant productivity. The old adage that if one gets things right from a human point of view then everything else can come right would seem to be well illustrated in this matter of worker/job match. In terms of humanity, productivity, and resource economy a very grave mistake is made when it is assumed that people must fit themselves to the job. The introduction of new microelectronic technology, which in product ion engineering provides the potential for a revolution in control systems design and worker/machine communication, gives design engineers an ideal opportunity to create production systems that make optimum use of both human skills and machine capabilities. The machine/ worker relationship will not only need to be considered within the narrow terms of production operation. The optimum performance of a machine/human combination depends to a considerable extent upon the amount of 'down-time' caused either by machine unserviceability or operator indisposition. The operation needs to be such that it increases neither of these causes of 'down-time'; and there is considerable evidence that where operators are responsible for maintenance and repair of their machines as well as for their operation, serviceability is at a high level and there is greater job satisfaction for workers. The achievement of an optimum machine/human combination obviously requires very close co-operation between operators and designers. Changing ingrained attitudes, both conscious and unconscious, is certain to meet with great resistance from design engineers. Nevertheless it is the most important thing that needs to be done in this area.

A systems approach

A drive for resource economy through manufacture in small dispersed units presents many interesting challenges to ingenuity. For many years past in Britain electricity generation has been a state monopoly, with only a small proportion of industrial consumption being satisfied by in-plant generators. Increasing fuel costs provide additional incentive and opportunities for the design, production, installation, and operation of 'combined heat and power' plants of various sizes from the smallest suitable for a single building, to the largest which might provide power and heat for a complete city. Between these two extremes there are intermediate levels of supply, such as the Fiat TOTEM system, which is capable of supplying a small housing estate, or the Hereford plant of 18 megawatt capacity, which provides electricity for the city of Hereford and process heat for the Bulmers Cider factory and the Sun Valley Products plant. Although research and development on power supplies for individual homes has been going on for several years, nothing suitable has yet been produced commercially. However, it has been reported that Rolls Royce engineers in Coventry have designed an engine, said to be the size of a shoe box, which could cut fuel costs in a typical home by about 25%. Run on natural gas, it is likely to cost about $400 to install and will supply all the heating, lighting, and cooking needs.

Once the overall thermal efficiency of power plants has been raised to about 75% from 35% by utilizing the low grade exhaust heat, electricity can itself be used with high primary energy efficiency to power heat pumps. Replacing a gas fired space heater or boiler by a heat pump can reduce energy consumed for space heating – which is much the biggest item in domestic energy – by a factor of three. So long as electricity is generated without waste heat utilization there is no thermal advantage in using an electrically driven heat pump; but when advantage is taken of the waste heat in a combined heat-power plant the full thermal benefit of the heat pump is obtained.

When space heating, water heating, and cooking are supplied by gas, the domestic requirement for electricity can be very small and will become even smaller as the potential improvements in efficiency of such important items as lighting come into use. For example, a low pressure sodium lamp is 15 times as effective per watt as a conventional tungsten lamp, and twice as effective as a modern fluorescent lamp. The very low level of electricity consumption in a household which incorporates such highly efficient lighting technology may well within a period of two decades be able to rely upon electricity supplied by its own photovoltaic system. The first practical solar cells were developed by the Bell Telephone Company in the 1950s. The investment cost per kilowatt generated was several million dollars. Since then rapid progress has been made in reducing costs and improving performance of cells. A plant is already in use in Saudi Arabia at a reported investment cost of little more than $10,000 per installed kilowatt, and Matsushita of Japan announced in 1983 a new type of cell for which the capital investment is $4,500 per installed kilowatt. At the 1981 Solar World Forum a subsidiary of Standard Oil of Ohio – Energy Conversion Devices – claimed that they were on the verge of producing solar cells at

$1,000 per kilowatt installed, which is very close to the US Department of Energy development goal of $700. Photovoltaic cells are no longer a scientific curiosity for the direct conversion of solar radiation to electricity. The first mass production plant, which started operation at Shinjo, Japan, in 1983, is a joint venture between Energy Conversion Devices Inc. and Sharp, the Japanese electronics company. The plant has an annual production capacity of 3 megawatts, and within five years there are plans to increase production to 120 megawatts per annum. Availability of photovoltaic cells in large commercial quantities will have a profound effect on many engineering developments in the coming years for both domestic and other small scale applications.

The need to obtain as much useful energy as possible from available resources has emphasized the need to consider the engineering of complete, and sometimes very complex, systems rather than merely seeking to maximize the thermal efficiency of a particular component in isolation. For example, the laudable aim of maximizing the efficiency of conversion in Central Electricity Generating Board power stations in Britain has resulted in waste heat from turbines at such a low temperature that it is unsuitable for space heating application. The adoption of a combined heat and power approach within the Central Electricity Generating Board, which will be necessary in order to make full use of the energy input to power generation, must have profound effects upon power plant design, both major central power stations and local stations. In the latter, consideration will need to be taken of the possible use of urban combustible waste as an energy source as an alternative to fossil fuels.

The availability of various alternative energy sources for power generation in rural areas presents a new range of engineering challenges. In addition to the engineering of total energy systems which can make farms self-sufficient, as described in Chapter 5, there is a massive potential waste biomass resource which needs to be 'harvested' and processed. The Energy Technology Support Unit at Harwell have estimated that organic wastes from all sources in Britain are approximately equivalent in energy value to North Sea oil production. Used either as a solid fuel or as a feedstock for conversion to methane gas or liquid fuel it could supply about 15% of the total national energy requirement. Most of this biomass is widely scattered in rural areas and imaginatively exploited it could make the entire rural population energy self-sufficient. In 1978 the estimated cost was about $30 per tonne of oil equivalent. With the exploitation of this important resource, combined with the use of photovoltaic cells and mini-hydro and wind machine developments, rural areas may no longer suffer the relative disadvantages of dependence on national energy sources of piped gas and centrally generated electricity. The engineering of systems designed to meet the needs of a dispersed rural population is as important for Britain as it is for the Third World.

A different kind of rural technology for the new conserver age provides another example of the challenges which face this and succeeding generations of engineers. Again it is small scale engineering. British Earthworm Technology, based in Cambridge, is exploiting worm-farming techniques developed at the Rothamsted Agricultural Research Station at Harpenden, Herts. Some worm species, such as the tiger worm, breed readily in a wide range of organic wastes – animal manures, sewage sludges, and domestic refuse. They feed mainly on micro-organisms growing on the waste material. In the process of feeding the waste fragments and is converted into rich nitrogen plant foods, whilst the worms themselves multiply as valuable natural animal feed for chickens, fish, and pigs. As much as 100 kg of worms, which as dry matter are 60-70% protein, is produced from every tonne of animal waste. Engineering is required for commercial worm extraction, and worm preservation methods, such as freezing, freeze drying, air and heat drying, and pickling or ensiling, need to be developed. To obtain the best yields and the most efficient waste conversion, worm farming requires a controlled environment. The systems that have to be designed require insulated beds, heat and moisture inputs, and ventilation control. A multiplicity of worm-farming plants throughout the country could replace some of the large scale synthetic fertilizer plant capacity and some of the conventional animal protein feed production. At the same time this technology could make a useful contribution towards the hygienic disposal of a great deal of organic waste, which in a modern highly urbanized society, requires massive engineering systems and heavy expenditure.

Simplicity

The most elegant solutions to engineering problems are characterized by their simplicity. They are usually also the ones which cost least. Primitive machine designs usually embody that quality of simplicity, but, as more and more is demanded of the machine in the course of its development, complexity sets in. A considerable amount of the complexity is a 'control gear', which may be mechanical, hydraulic, or electrical, or a combination of two or three types. Most of the control gear is a substitute for the manual control of the human operator. It is not surprising that any attempt to reproduce the sensitive mechanisms of human manipulation is complex and costly. It may well be that a radical new approach to design, in which a better match between machine and operator is the aim, will of itself recover some of the simplicity that has been lost as a result of 'machines designed for morons'. It is also very likely that a degree of simplicity will be achieved through the application of microelectronics to the control gear of machines.

Small scale hydroelectricity generation provides a useful example of the simplifying effect of electronics control when compared with the mechanical gear that was used in Victorian machines. In the nineteenth century small water turbo generators were to a considerable extent scaled down versions of major hydroelectric systems, using the same principles of mechanical control. In the scaling down process the control gear became a much greater proportion of the cost of the total system, making the capital cost per installed kilowatt much higher. With the passage of time, and with the increased availability of alternative sources of electricity at lower cost, small water turbo generators ceased to be used. With the development of electronics a new approach to design of small plant has been adopted by the Intermediate Technology Development Group. The old costly mechanical control gear has been replaced by an electronic load controller. This has reduced the capital cost of small scale hydro-power plants to such an extent that they are once again a sensible investment even in poor Third World rural areas. The load controller design is simple enough to allow assembly in rural workshops.

Other work done by the Intermediate Technology Development Group has demonstrated that further simplifications of traditional design practice can be achieved when simplicity is made a mandatory feature of a design brief. In their work to develop wind-powered pumping machines for small Third World irrigation projects, the aim was not only to produce a design which would be much cheaper than conventional wind-driven pumps but also one which would require minimum servicing and be capable of manufacture in simple rural work shops. In a conventional wind driven machine the head mechanism, which converts the horizontal torque to a vertical shaft, is a heavy and costly gearbox assembly. In the ITDG design the gearbox is replaced by a simple link mechanism which is only a fraction of the weight and cost of a gearbox. Because the head assembly is so much lighter it is possible to use a lightweight tower, with further savings in weight of material and cost. By incorporating sealed-for-life bearings and self-lubricating bushes no lubrication service is required and breakdowns are avoided. Similar experience has been gained by the Philips Electronics company in Utrecht when they have set out to develop special products and manufacturing processes for use in their Third World factories. They have often been surprised by the ingenious solutions that have been developed when simplicity was a prime objective; they have been able to apply the lessons learned from this work to their European operations.

Not only can ingenious simple engineering solutions be produced by taking a second look at conventional designs with the intention of simplification, it is sometimes possible to recover simplicity by reviving old technologies. The Humphrey pump is an example. Invented by a very ingenious British engineer, several very large units were installed in England early in this century. The water to be pumped is directly exposed to the expanding combustion gases in a cylinder, and itself acts as the piston driving the column of water into a high reservoir. It is extremely simple to make, there being no need for precise machining of cylinder and piston, it requires no lubrication and suffers no wear. In recent years Reading University and the Intermediate Technology Development Group have revived interest in the Humphrey pump. Various sizes have been designed and some field testing has been carried out in various overseas countries. One example capable of pumping 225 metre^3/hour to a head of 9 metres, has been field tested in Egypt by the Egyptian National Research Centre. The application of many of the lessons learned from the development of 'appropriate' small scale and simple, low-cost technologies for Third World countries can provide an invaluable reverse technology transfer as Britain, and other industrially advanced countries, change direction of engineering for a conserver economy.

Scale

For more than three decades most technological developments have tended to favour large scale applications. In recent years many developments have given considerable encouragement to the recovery of small and medium scale operations. Progress in electronic engineering and computer technology has been extremely rapid in the past two decades; and applications which only a few years ago were economically possible only to major industrial plant are now available to very small plant at a fraction of the earlier cost. As a result of these and other developments much shorter production runs are now possible with profound effects on product design and the whole manufacturing operation. Nowhere have these developments had a more dramatic effect than in the printing industry.

This technological development factor, reinforced by a growing concern about the failure of many large scale projects to yield expected economies and other advantages, has promoted a critical reappraisal in many different industries of the most effective plant size in relation to the size of market to be supplied. As a result of that reappraisal there has been a gradual retreat from the megasize plants that were fashionable in the sixties, across a broad spectrum from oil tankers to breweries, from steel plant to bakeries.

There is another way in which engineering and technology developments are having an effect on the scale of industrial operations. The development of the Fairford ac induction motor electronic controller, in which ITDG played an important part, provides a dramatic example of this effect. Three phase ac induction motors consume almost two-thirds of all the electricity used by British industry and nearly one-third of the entire output of the national grid. The Fairford controller, which is based on a modern digital electronic circuit incorporating a microprocessor, is capable when fitted to each of the many thousands of industrial electric motors of saving about 15% of the electricity they consume. In other words this small electronic device, when applied widely enough, is able to displace several major power stations and the coal mines, the oil fields, or the uranium mines that supply their energy.

CHAPTER 6
Being your own boss

The importance of new firms

A virgin forest consists of a complex mix of vegetation. Large trees stretch out their canopies of leaves to the light. Beneath them smaller trees and bushes take root, growing up towards the roof of the forest. Under these lies a thick tangle of undergrowth – leaves, creepers, and shoots; this undergrowth is the nursery where seedlings germinate and life begins. A sustainable ecosystem needs a careful balance of vegetation: a few big trees, many smaller ones, and thousands of seedlings. The big trees provide the humus and an umbrella to shelter the new growth, and in return the seedlings cloak the ground, retaining moisture for tree roots.

'Large streams from little fountains flow, / Tall oaks from little acorns grow', wrote David Everett in the eighteenth century. Oaks and acorns need each other equally, but at the same time there is healthy competition between them. Most acorns fall on barren ground, or do not germinate, or their shoots remain small and stunted. Only the strongest stock grow tall and healthy, eventually to replace the old enfeebled trees in the forest canopy. Clear the undergrowth and there are no new seedlings to replenish the forest. Big trees grow old and eventually die, but the system cannot regenerate.

So it is with the economy. Just like a forest, an economy needs a balanced mix of large and small firms. Like the oak and the acorn, large firms can help to nurture smaller ones, while the latter keep the former awake and competitive. One day a few of these smaller firms will grow up to replace the ossifying older ones, bringing with them new ideas, new technologies, new management, and new vigour. Only in this way can an economy rejuvenate and adapt to a changing business climate in the world.

That is why the undergrowth of the economy is crucially important. This chapter looks at the new seedling firms that prosper or decay here in Britain, who plants them, and why they are reviving.

The decay and revival of small firms

Until the industrial revolution almost all industry in Britain either took place in the home or was done by specialist craftsmen and tradesmen in the village. Take bread as an example: wheat was ground into flour by the miller, then baked into loaves in the oven at home. It was only when people started to move off the land and into towns that the earliest commercial bakeries were set up. No longer could you walk to the mill and bring home a sack of flour. With less time and more money on their hands, the populace began to buy their bread.

By the end of the nineteenth century only 10% of the population was left working on the land. It was the age of the small business in the town and the city. Baking was no exception: by this time there were thousands of bakeries through the land turning out all sorts of breads from their wood-fired ovens, depending on local tastes, some of it good quality, some of it bad.

The twentieth century brought changes. Some of the bigger bakeries began to expand, buying up smaller ones and closing down their ovens forever, but using their high-street shop fronts to expand sales. Their bread was baked in large modern ovens in a central factory, to be delivered each day to the shops. The food regulations of the Second World War helped the larger firms to strengthen their stranglehold over the industry, and many small firms never started up again afterwards.

The bread itself was changing too. In the laboratories of the big companies scientists developed a standard white loaf, bland, soft, and fibreless, a loaf that would taste the same from Scotland to Cornwall and that would keep for weeks. Consumers bought this white factory bread, and all over the country small craft bakeries closed down. Where once there had been thousands, by 1970 only 300 bakeries remained.

So it was with other industries. Everywhere small firms were being taken over by larger ones and closed down or used as marketing outlets. Production moved from local workshops to factories, where craft work was forgotten and mechanized techniques of batch or mass production were used to turn out standardized goods for a regional or national market. The countless small manufacturing firms in the nineteenth century diminished until by 1930 there were only 150,000, and by 1970 fewer than half of these remained.

A low point was reached in 1970. Relative to its size Britain had fewer small firms than any other major country in Western Europe or North America. Furthermore, these small firms offered fewer jobs and contributed less to national production than in competing countries. The undergrowth in Britain's economic forest was looking dangerously thin.

At last the government stepped in. Worried by the rapid decay of small businesses, they set up a weighty commission of enquiry to investigate what was happening. This resulted in the Bolton Report, a number of volumes of evidence that spelled out a clear message; the small firm sector was in danger, and something needed to be done about it fast.

Why worry about them? Because, said Bolton, far from being the archaic and inefficient survivors of an earlier age, small firms were crucial to the efficient functioning of the modern economy. Small firms, they pointed out, provide an outlet for entrepreneurial talent, a seedbed for innovations, a breeding ground for larger firms; they offer the most technically efficient way of producing certain goods and services, and they provide variety, choice, and competition in the economy. Had they known what we know now, the report might also have said that small firms provide a crucial proportion of new jobs.

The Bolton Report marked a turning point in the fortunes of the small business. Today public attitudes are changing, as too are government policies, the views of large firms, technologies, and trading conditions. The age of small firms is returning, albeit in new forms and filling new roles. Since the low point of 1970, about 16,000 new firms have been set up every year in the industrial sector alone. We look at this revival below.

In 1975 a young man named Robert Gordon plucked up his courage and took the decision to leave the big bakery where he had worked for some years in order to set up on his own. Striking out in this way seemed a very individual thing to be doing, but in fact Gordon was on the leading edge of a national groundswell. He took over an old building that had once been a bakery in a small Lake District town, and after some renovation work to allow for new equipment and to meet the requirements of the health inspectors, was soon stoking up his ovens ready for his first baking.

Today his bakery is prospering, producing wholewheat breads, quality pies, and traditional cakes. If you ask why he doesn't bake white bread, he answers that the large bakeries can do that best, and he would rather produce the goods a small craft bakery is best at. Robert Gordon is not alone: several hundred small bakers have set up in the last few years, specializing in fancy breads, special flours, and regional products. The industry analysts say their prospects look good and this success is likely to continue.

What has happened in the baking industry has also shown itself in other areas of the economy. It is a movement that is attracting people from many different backgrounds, drawn by the prospect of being their own boss for the first time. Why should they be tempted by this prospect?

Why people start new firms

Of the work-force in Britain today, almost all are employed by a private business, a corporation, or a government department. Fewer than one in ten work as their own boss, but this proportion is increasing. In later chapters we examine why small firms may have a better chance of success than in earlier days, but in this chapter we are interested in why people want to go through the gambles, hard work, and setbacks of starting new firms.

There are a myriad reasons. Economists traditionally assumed that the main one was the opportunity to make money. Small business was an outlet for those who saw a gap in the market and wanted to cash in on it. This is still true, but it is by no means the only motivation today. Indeed, the rules of the game are such that it may now be easier to make money through investments in the corporate sector than by setting up a new business. We look at some of the other attractions.

One prime motivation is dissatisfaction and boredom with work. Many founders of small businesses worked for other firms for years without feeling fulfilled. Some were bored as craft-based work gave way to de-skilled jobs on the assembly line. Others found that working for a big organization was impersonal, or they felt their work was not appreciated by the employer, or that they could organize the business better themselves. Most employees have felt this at some time or other. While some swallow their pride and press on, those that do not are one type of new firm founder.

Will Peters is one of these. His family had a share in an old small brewing company. When the time came to leave school Will did as was expected of him, following his father into the firm. By the time the older man had retired Will was well established in the business. Then the little company was bought out by one of the giant national brewers. The new owners were interested in the public houses that the family brewery owned as outlets for their own beer, but the brewing plant was considered antiquated and was closed down. They offered Will a job in one of their modern large scale automated brewing plants. But try as he might, Will could not fit in: he argued with his supervisor, he did not like the beer he had to brew, and he did not like being a white-coated technician in a sterile plant rather than a craftsman brewer in gumboots. One day Will arrived at work very late, then he was rude to his supervisor. Not long after, he received a message from the company giving him notice to leave. He was, they said sadly, unemployable.

Will Peters is now a rich man. The company was probably right when they called him unemployable. But far from feeling downcast Will felt his chance had come to do what he wanted – to return to craft brewing.

He thought about the market, and he looked at small scale equipment, and he talked to many people inside and outside the industry. Most of them laughed at his ideas. Being pig-headed, he ignored the warnings and went ahead on his own, borrowing money and setting up a tiny brewery to make short brews of strong traditional ale. Ten years later, after many setbacks, he is a very successful business man, still brewing these ales. His strong will and determination, precisely the things that made him such a bad employee, were what helped him to succeed on his own.

The economist Schumpeter, who was one of the first to look at new businesses and their role in powering economic growth, took the view that many firms were set up to exploit something new: new ideas, new processes, new technologies, new products, new markets. People pick up bright ideas at their work, in their leisure interests, or about the house. Because it is hard to get firms already in the market interested in taking on new ideas, one option is to set up a business to carry out the idea oneself.

This path is a risky one. A business idea does not necessarily make a business, as we shall see; many of the best businesses come from replicating very ordinary ideas indeed; and a person with an inspiration may find it very difficult to analyse the idea critically and in commercial terms. For every Clive Sinclair who succeeds, hundreds fail. Nevertheless this is an important motivation for new businesses, as a middle-aged couple in Wales can tell.

Visiting plant nurseries they realized there must be a commercial future for packaging young plants in such a way that they could be sold through non-specialist shops. After much thinking they developed a simple process that compacted shredded paper that had been used as chicken bedding, and so contained useful nutrients, into tray-mounted pots. Seedlings could be planted in each pot and a shop would buy a tray. Because the pots already contained moisture and nutrients the plants needed no maintenance. When a customer bought one they simply cut the pot off the tray. A simple idea that led to a successful business.

However, not all businesses start in such a positive way. For more and more people it is not a creative new way of working, but rather one forced by cold economic hardship. As jobs have got scarcer and dole queues have grown, the unemployed and people fearing they might soon be unemployed have looked at the possibility of self-employment. This is perhaps the strongest motivation of all. Such new businesses are usually very small, with minimal capital, and they often concentrate on providing services to households or to other firms. Some fail, but others provide a livelihood. A common problem is that unemployed people find it hard to finance the time it takes to get a business started. The government's Enterprise Allowance makes this a little easier however, and is helping the revival of one-person service and craft industries.

Travers Sharpeners Ltd started life this way. Ted Travers is a cheerful middle-aged man who used to work in an engineering shop. In the late 1970s things started going badly for the business, and before long Ted was made redundant. To make things worse, his wife died suddenly, leaving him with two young children to look after. Ted faced up to the facts: he needed to earn some money urgently, but it had to be on a job with flexible hours so he could be at home when his children returned from school. After a lot of thought and a few enquiries around the neighbourhood, Ted got hold of an old van and rigged up a grinding wheel inside. With this basic equipment his little sharpening and grinding business was born. Travers Sharpeners moved around the towns in the area sharpening knives for restaurants, scissors for hairdressers, and tools for hardware shops. Business built up as his reputation spread and regular customers asked him to return. Today it is a flourishing little business that cost Ted plenty of work but only a few hundred pounds in capital.

The new starters

Who are the Ted Travers and Will Peters and Robert Gordons of this world? Are they born business men and women, or can anyone do it?

The old view of entrepreneurs was that they were not just hard-working and clever, but also that they were foolhardy gamblers, social misfits with a grudge against society and a determination to prove themselves right. 'Are you a heavy drinker,' asked an article in the Guardian ten years ago, 'a compulsive gambler, a womaniser, a reckless driver?' Only if you filled these and other requirements were you statistically likely to be a successful business man.

Today's entrepreneurs, however, cannot be thus characterized, because they include a whole spectrum of men and women. Sometimes they have a business background, sometimes they are manual workers or professional people moving into new fields. The rules are changing, and the new starters are people who might never have considered business before.

Early retirers are one type of new business person; men and women who have spent their working lives as employees but who have some special interests they are keen to develop. When they retire they have time to pursue these and look at their commercial potential. For example, a glance at the wood products industry shows thousands of craftsmen turning out bowls and toys, picture frames and earrings, musical instruments and wooden components. Many of them first began to make these things as hobbies.

Women are becoming more involved in business. Most of their businesses result from commercial applications of traditional women's activities such as cooking and dressmaking, often aimed at female customers and sometimes started on a part-time basis from home. Increasingly, however, female-run businesses are to be found in non-traditional sectors of the economy. One industry which they have always dominated is craft knitting. Here there are women entrepreneurs and collective groups who produce machine knitted jerseys, hats, socks, and leg-warmers for boutiques and markets, as well as fashion garments for special orders. For those with the rather rare combination of business acumen and design skills that is required, there can be a good living here.

There have long been Asian business men in the UK. Now other minority ethnic groups are also starting businesses, some to meet demands for exotic or ethnic products and services, others aimed at a more general market. Support organizations that have been set up to encourage these minority groups into business report on growing successes from bakeries to furniture makers.

Young people, too, are looking at the possibilities of business today. They face the biggest problem of all. It is a lot to expect school-leavers with no experience of work, let alone of commerce, to become entrepreneurs. Nevertheless, the new times have thrown up many stories of youngsters who form software firms or computer maintenance companies while still at school. Others have set up groups like Instant Muscle; this is a collection of boys who, lacking any specialist skills, offer to do any sort of manual work around the community. The original group in Surrey has been so successful that a number of others have now set up around the country.

Ideas for business

Where do all these people, the successful and the unsuccessful, get their ideas from and how do they develop them into businesses? A small number come from bright and unusual ideas such as the Welsh plant-pot makers, but 95 out of every 100 successful businesses result from their founders spotting gaps in particular markets, then copying another business with some modifications.

A large firm would not consider new product ideas that did not fit in with its image, its production methods, and its marketing structure.

Similarly, an entrepreneur should think twice about a serious mismatch between his or her skills, experience, and resources, and an intended business. For those without direct commercial experience it is inevitably difficult to identify appropriate opportunities, but ways do exist.

One approach is to identify where it has proved hard to buy some product or service, then consider the possibility of a business to provide it locally. Another is to look at new products that are advertised with a view to distributing them locally. New trends such as a rising tide of vandalism or changing clothes fashions can throw up business possibilities. Legislation imposing new regulations on industry or commerce may present opportunities. So also can changing economic circumstances such as the energy crisis.

A stroll through a museum reveals many long-forgotten products and pieces of equipment, some of which may have a practical or curio value today. Department stores and supermarkets, especially the more innovative ones, stock many products that could be produced on a small scale. A trip abroad with an open-minded wander through local shops and markets can be very fruitful. Trade fairs and business exhibitions are similarly full of ideas for the industrial and commercial markets.

Other business leads and potential competition can be checked by reading through the local yellow pages of the telephone directory, trade directories, and specialist trade journals. The newspapers, especially the business pages, are full of items that can throw up ideas, while the 'businesses for sale' column and advertisements for inventions seeking capital are also useful.

But an idea is not a business. The next step for the would-be entrepreneur is to translate the good business idea into a commercially viable enterprise. This is likely to be a lengthy process with numerous setbacks. The idea first needs to be teased out to assess the potential for a product or service that can be sold. Then it is time for research to decide where the market is, its size, the technical requirements of the business, and the costs and returns that might be achieved. If the business idea needs to be modified at any stage, this series of steps is retraced.

At this point the real work starts. If the original idea holds up it is time to start experimenting with sample work to test out the market further, on the basis of which the original idea may need to be recast or even totally rethought. Then come all the tasks like finding premises, buying equipment, obtaining finance, taking legal advice, and, finally, starting up. This is a long process. Most new firms are in this transitional state for at least a year before they open for business, and before that they may have been a dream in their founders' minds for many years.

The early days

To return to the analogy of the forest, the seed has now dropped to the forest floor, where it is germinating. If the climate is right, it will take root and start sprouting. But if the ecology of the forest is not stable and balanced, the seedling may die and in the long term the mighty forest itself can suffer.

In the past most small businesses were left to fend for themselves on the forest floor of the economy, but increasingly they have the opportunity to start up in an environment that will shelter them from the harsh winds of the outside world. Just like seedlings in a plant nursery, they have a better chance of survival here.

There are many sorts of sheltered nursery, as a few examples will show. Will Bright, an inventor and engineer, designed a prototype low-cost wind generator for use in Third World countries. From his experience of industry he realized the only way this generator could reach the market would be to manufacture it himself. But Will had had no experience of either batch production or business management. Consequently he took up temporary quarters in a managed workshop, one of many that have been set up in Britain over the last few years. Here he pays low rent on a short-term tenancy without having to commit himself to long-term obligations. He also has access to the workshop's machinery, so he need buy only the minimum at this stage; professional help on cash flows and bookkeeping is close at hand. Thus Will is spared from having to invest heavily until he has completed his first batch of 50-100 generators. By then he will know a lot more about production and marketing, and he will be in a strong position to move into the outside world.

Another way to shield new start-ups is by using a co-operative. Here a group of workers take shares in their business and work together, thereby avoiding the loneliness of the small business start-up and sharing some of the risks. Despite the failure of a number of older and larger worker cooperatives, this form of organization has become extremely popular in recent years. In the 1970s a number of co-operatives were set up in fields like printing and publishing, wholefood catering, and computer software. In the 1980s this broadened considerably and now co-operatives are to be found in many sectors of the economy, with new ones starting up all the time.

In London in the 1970s a group of design students were attracted by the good design and functional potential of traditional Japanese furniture. This consisted of stuffed cotton mattresses that folded on frames for use as beds, chairs, or sofas. Here, they mused, was something suitable for small flats in Britain. After considerable discussion and work they redesigned a few pieces into items that they could manufacture and sell. They formed a co-operative so that all the workers had an equal stake in the venture. They did very well, founding first a factory and then several shops, and are planning new co-operatives to produce their furniture in other cities.

The franchise is another form of business that allows a sheltered start-up. After paying a franchise fee a new starter receives very detailed commercial advice, a business plan, financial assistance, free advertising, and help with premises and equipment. The franchisor helps the starter to set up a local business on the lines of a tried and tested model. The advantage is that relatively few properly franchised ventures fail. The system works best for businesses that serve a confined local market and therefore can be replicated in different locations. Examples are fast-food shops and drain clearing services.

Some franchises, it must be said, are poorly run, and most are expensive. The Intermediate Technology Development Group is looking at ways to establish an 'ethical franchise' system whereby low-cost businesses can be replicated cheaply. This is likely to attract unemployed people who otherwise could never have afforded to buy a franchise.

A tree that drops seeds to the ground may shelter them from rain, wind, and predators with its overhanging branches: a large firm, too, may sometimes provide shelter for a small business that spins off from it. This has been much more common in the US than in Britain, but over the last few years, many large British firms have slimmed down, and some have helped their redundant ex-employees to go into business, sometimes contracting out to them work that was previously done in-house.

Whitbreads, the brewers, now employ small groups of ex-employees to do some of their wholesaling operations. An ICI plant has helped its former gardener to set up a business, which it now contracts to maintain lawns and flower borders. Rank Xerox have a large experimental scheme whereby redundant employees are assisted to set up businesses with the Xanadu Association, funded by the company, which helps with premises, services, accounts, marketing, and advice. Those who were computer users have been given their office equipment to take home, from where they keep in computer contact with Rank Xerox. The company assures them of full- or part-time work for a period, giving them time to work on setting up other contacts and getting other customers. After a year all 120 new starters were still in operation.

The role of the underground

Back to the forest: some plants survive their early days only by burrowing underground and establishing networks of roots, before pushing their shoots above the earth. It is the same in the economy. A favourite, though sometimes illegal, way to start a business is in the black economy. Here an operator can commercialize a hobby, try out a new business, or practice a craft, without getting too involved in costly start-up expenses.

If the climate is too severe the plant loses its new growth, but the roots remain intact beneath ground. This also happens in the underground economy which is burgeoning in Britain and other European countries today. One motive for the underground economy is to avoid paying taxes or fulfilling safety and other legal requirements; this cannot be defended. However, another motive is to experiment with new businesses; in this the underground economy fills an important adjustment and restructuring role.

In the West Midlands, cutbacks at car assembly and auto-component plants left a number of skilled and semi-skilled workers without jobs. Since then spare parts distributors in the region have reported unexpected sales of car spares. What has been happening? Some redundant engineering workers have been filling their time by overhauling their cars, and then by repairing the cars of friends, neighbours, and relatives. For some this has become almost a full-time job, working on the kerbside with borrowed tools. In future years some of these mechanics will hire proper workshops, buy equipment, and set up as proper garages, rejoining the above-ground economy. For them, working underground will have fulfilled its purpose.

Will new businesses succeed?

When botanists study the organic patterns of how trees live and die in a forest, they find that millions of seeds fall from the branches, or drift in on the wind, and bury themselves in the humus of the forest floor. Of these millions a relatively small number germinate, sending out roots and shoots. In the struggle to survive and establish a foothold only small numbers succeed. It is much the same with new firms. Many people have ideas about starting a business, but the number that are actually established is far smaller. Even then there is a painful weeding-out process, as the more ill-conceived or unlucky ventures fail.

The start-up period typically takes six months to three years, and heavy costs are undertaken. Money is needed for premises, equipment, labour, materials, and working capital. During this period the business starts to market its product or service, but there is likely to be a net loss on operations. There is also a high likelihood of failure at this time; one in ten of registered businesses fail in their first year, and one in three in the first three years. In fact the true failure rate must be even higher, because very small businesses do not have to register. This may sound altogether too risky. However, it should be pointed out that those business men and women who take training courses and obtain proper counselling and advice can reduce these risks considerably and have a much better chance of success.

The second stage of the business is the period of early growth. Markets are established, production is geared up, and new staff may be taken on. This phase can last several years, and during it a net profit must be achieved and debt repayments started. There is still some danger of failure in this period, with another 10-20% of firms going out of business. A common failing is the mistake of trying to expand too fast.

Darwin explained how, with plants as with all species, nature weeds out the weak or poorly adapted in order to ensure the long-term survival of the forest. As the climate changes, species must adapt themselves to it. In the same way, business failures are the economy's way of weeding out the unfit, ensuring a healthy stock of businesses, and helping industry and commerce adapt to change.

During the next phases of consolidation and maturity the enterprise is geared up to the most efficient scale and operating routines are established. The chances of failure are much more remote now, but as with the trees in the forest, there is always the chance of unexpected disasters: strikes, exchange rate movements, oil crises, industry contractions, credit squeezes, client bankruptcy, and legal changes could all put businesses at risk. The consolidation and maturity phase may take a few years to achieve or it might take a decade. It is only now that a reasonable return on the years of investment starts to be achieved. Consolidation does not necessarily mean growth. Most small firms will never grow, either because their founders do not want it, or because they are more efficient on a small scale.

New firms and the economy

Why worry about these small firms? Because they add up to an important source of jobs and of dynamism in the economy.

In 1979 a study by David Birch of new firms being set up in the US surprised many people by finding that two-thirds of net new jobs over the past decade had been created by firms employing fewer than twenty workers, a considerable proportion by new business start-ups.

Subsequent research in Britain has not shown such a strong small firm contribution here, partly because there are far fewer small firms in this country. It seems that important employment generation in the UK is coming from young firms rather than necessarily small firms. But certainly the contribution to employment has been most noticeable in regions like the west Midlands where the base of small firms is strongest.

Just because more and more small firms are being set up it does not necessarily follow that unemployment problems are going to be solved. Over the last decade small firms have doubled their share of employment, but the fact remains that it would take hundreds or even thousands of new small firms to repair the damage to employment caused by the closure of one big factory. Small firms do not yet provide enough places to reduce unemployment, but a job-seeker is relatively more likely to find work in a small firm rather than a big firm today.

The other main contribution of small firms lies in their ability to invigorate the economy and provide the elements needed for change and choice. Small firms help an economy to remain dynamic by spawning new ideas, testing commercial possibilities, adapting new technologies, providing outlets for the energy of entrepreneurs, and acting as seedbeds for the talent of investors. It is no coincidence that Britain, with few small firms, has had more trouble in adjusting to post industrial structural changes than have most of our competitors.

The growth of large companies and concentration of ownership in a few hands has cut down the choice of products and services available today. In many industries there are only a few national brand names, which may be essentially the same, to choose from. It is up to small firms to maintain local products, customized services, special jobs, and in doing so to keep the big firm on its toes. Small firms can usually produce short runs and do jobbing or customized work more efficiently than the large firm: Chapter 7 looks at this in detail.

The importance of small firms has now become more obvious to the public, and some of the fog of misunderstanding and prejudice which has surrounded them in the past is at last starting to clear. However, there is still much that stands in the way of small businesses; these barriers form the substance of Chapter 8.

Reference

Birch, D. (1979) 'The job generation process', MIT Program on Neighbourhood and Regional Change.

CHAPTER 7
Big prospects for small firms

Big is bountiful

Today there are few car plants left in Britain but not many years ago most sizeable villages had their equivalent of the modern car plant: cart-building workshops. To make a cart required a carpenter to build the body, a wright to make the wheels, and a smith to do the ironwork. All highly skilled craftsmen, they used tried and tested techniques and the result was effective enough by the standards of the day. The traveller can still see similar carts being made in similar ways in villages all over the Third World, but British villages have lost their ability to build their own transport. Instead there are a tiny number of huge car plants. In the technical sense at least, this represents progress.

The craftsmen of earlier generations, forerunners of today's industrial manufacturers, had little choice in the techniques they used, and were able to produce only on a very small scale. Over a number of years, in a slow and sometimes painful way, the industrial revolution changed all this. New techniques, new materials, the application of science to industrial processes, and the availability of commercial credit meant that certain operations could be done more efficiently on a larger scale. The industrial factories of the north bore witness to this change, first with coal mining, steel, and textiles, and later with other industries. By the twentieth century, systems for handling bulk products, the national transport network, the building of warehouses and improved stock control, advertising and national marketing, and the refinement of techniques of business management all reinforced the trend to large scale production. When Henry Ford built his car plant based on the new concept of mass production he exploited all these techniques to produce cars cheaply. How far the industrial system has changed in Britain today is illustrated by the car industry. Instead of the cart-maker in many villages, there are now only four main plants making British cars. They are highly automated and extremely capital intensive, using an ever smaller work-force to turn out vast numbers of cars. Whereas the pre-industrial craftsman used the equivalent of a few hundred pounds' worth of equipment to build a cart every few weeks, the average car worker uses equipment worth more than £20,000 to build a sophisticated car in one or two weeks.

A walk through Longbridge, the British Leyland plant outside Birmingham, shows how far the industry has come. The visitor stepping off the train at Longbridge Station sees the factory across the road but, as if to emphasize that this is the age of the car and not of the train, the visitor's gate is half-an-hour's walk away on the far side of the plant. On the way one passes the huge stores buildings, the clattering canteens, the acrid-smelling paint shops, the clanging assembly plants, and the spreading acres of parking space for the vehicles that roll off the lines each week. Inside, men and women work, computers check the running of the system, and robots weld and spray. To the visitor the plant looks vast and unmanageable. To the manager the worry is the reports of his analysts, who say that Longbridge may still be too small and too old-fashioned to survive in the modern world of auto-making.

In the 1970s this trend to giant plants was justified by the phrase 'economies of scale'. This means that the unit costs of production can be cut by increasing output. To make one BL Metro might cost £20,000, to make hundreds could cut the cost to £10,000 each, while to make thousands could reduce unit costs to only £4,000. This principle was considered to apply to the making of goods in the factory, to the management of the business, and to selling on the market. The mood of the 1970s was that economies of scale applied to most types of production, and plants that were too small would have to be 'rationalized'.

Economists use the term 'minimum efficient scale' to indicate the importance of large scale economies in an industry. Minimum efficient scale is the smallest size at which a plant can operate without incurring significantly higher costs. The brewing industry was thought to be an example where big might be bountiful: in fact so rapid were developments in the industry that analysts had to keep revising upwards their estimates of minimum efficient scale. In 1960 it was estimated that the smallest efficient output for a brewery was 60,000 barrels per year. By 1971 this had been revised upwards to 1,000,000 barrels, and by 1980 two new plants had been built with a capacity of over 2,000,000 barrels each. It seemed that modern plants must be ever bigger and bigger.

From this situation it was a small mental step to thinking that 'large scale' necessarily equated with modernness and efficiency, and that small scale must therefore mean old-fashioned and inefficient. This thinking permeated the minds of industrialists as they ordered new plant, it lay behind the thinking of civil servants as they implemented 'rationalization' policies, and it filtered through the attitude of the general public as they decided how to invest their savings and where to buy their goods.

The 'small is beautiful' revolution

Just as the industrial revolution and its aftershocks upset the industrial balance of the eighteenth century, so the same thing is happening in a small way today. For want of a better name this has been called the Post-Industrial Revolution, the cause of the big changes in work and employment that we have been discussing. Less evident, but equally important, is the fact that the post-industrial revolution is also changing the balance of industrial power in the economy.

These changes are not just short-term jolts that last for a year or two, like a recession. Rather, they are long-term trends which take some time to show their effects but have far more impact over a period. Such long-term economic changes have been documented for years. The 'Kondratieff wave' is one such cycle, postulating long periods of innovative activity and growth followed by long periods of decline: 1940-75 was a period of growth; 1975-2000 is thought likely to be a period of decline and structural change as new methods of production come into use.

Evidence of this trend is seen in the changing pressures on industry today. We look at three main areas in the rest of this chapter. Firstly, there are new and adapted technologies available today that are changing the techniques of production just as the industrial revolution did. Secondly, there is a changing business climate resulting from the energy crisis, pressures on finite resources, increasing competition from newly industrialized countries, and a decaying infrastructure in Britain. Thirdly, there is a change in the attitudes of people inside and outside industry to the relative roles of big and small business. There is a realization that the same big plants which can generate savings through manufacturing large orders, are also open to a host of new problems that small firms escape, and the whole question of economies of large scale and economies of small scale is now undergoing radical rethinking.

Changes in industrial technologies

The post-industrial era has brought with it a range of new technologies, some of them radically innovative, others merely adaptations or new applications of well-known traditional technologies.

Microelectronics

The first industrial applications of electronics were based on large mainframe computers. These offered new methods of stock control, operations analysis, and data storage, allowing improved management of large scale operations. On the factory floor they allowed for centralized control of machinery: banks of numerically controlled machine tools could be fed instructions and have their performance monitored by an office-bound controller. These computers were hugely expensive machines, specialized in use and cumbersome to operate, and only large firms were able to make use of them.

By the 1970s developments in microprocessors were bringing mini- and microcomputers on to the market, prices dropped drastically, and as they did, microelectronics-based control systems found applications in many items of production equipment. Now the microcomputer is cheap enough for the smallest business and specialist software has become available for a wide range of small scale uses, from book-keeping for hairdressers to metering for taxi-drivers. With computer-aided design systems and add-on direct control units for machines, production control has been brought back to the individual operator on the shop floor.

The microelectronic revolution has not yet finished. While most attention has focused on its effects on jobs and skills, it has also brought sophisticated control systems within the reach of the smallest producer. One of the earliest industries to benefit from microelectronics was the printing industry. Britain has an old and highly skilled industry which consists of a few large newspaper and book printing companies, and a myriad tiny jobbing printers who take on small orders, from menus to church magazines. Printing presses have developed over the last century and are relatively efficient. But composition, that is, providing the type or plate from which the printed impression was made, remained rather archaic. Compositors sat at heavy machines punching out slugs of metal type which were made up into blocks by hand. This was an obvious area of application for electronics, and by 1970 photocomposition techniques had become common. Now the operator could sit in front of a screen, type in text, correct and space it automatically, then store it on disc or tape ready for plate-making. The early systems were developed for the large newspapers with banks of operators, and were very fast but expensive. However, because there are so many jobbing printers and they represent an important market for equipment, the manufacturers were not long in developing photocomposition systems that were much cheaper, smaller, cleaner, quiet, and more flexible, filling the needs of the small business. Now the small-town high street printer whose bread and butter work is printing invitation cards to the local wedding and producing the parish magazine can and frequently does use equipment which is as advanced as that found in Fleet Street.

Printers have been lucky; in other industries technical improvements have been slower and small firms less well catered for.

Automation

While the microelectronics revolution is now well under way, developments in automation promise another revolution, but its likely effects are far less clear. Automated systems allow machinery to carry out advanced mechanical operations such as loading, machining, checking, and stacking under their own controls. Machining centres that can carry out a range of operations are one form of automation already found in some engineering shops. Robots represent another that is on the way. In a small number of very large plants there are already lines called flexible manufacturing systems, where a component can move from warehouse to assembly centre, be manufactured into an article, then packed for delivery, without direct human help.

Automation in manufacturing seems likely to reduce the cost of small batch production. But this is the traditional preserve of the small firm. So who is going to benefit from automation? Will traditional small firms be able to use it to upgrade their jobbing work, or will the benefit accrue to the engineering giants who have the resources to develop this technology in their own image? At this stage the latter seems more likely. Advanced machining centres cost in the region of several million pounds, far beyond the budget of most small firms. Under the Automated Small-Batch Production scheme, appropriately named ASP, the government is paying large sums of money to induce a couple of big firms to develop further sophisticated flexible manufacturing systems that small firms will never use. The next step is the unmanned factory: the Japanese have already built a show-piece factory to manufacture robots, which runs with no-one on the shop floor. In Britain, a show-piece factory in Colchester, largely funded by a £3 million government grant, produces engineering components with only a handful of white-collar operators to switch on.

However, not all developments are leaning in this direction. Industrial robots are still in their earliest stages of development and are very expensive precision instruments. So far the jobs they have replaced in industry have mainly been dangerous, polluting, or repetitious ones such as spray painting, welding, and packing. When coupled with remote control techniques they make it possible to do work in inaccessible places, such as non-man-entry sewers.

There are also signs that much cheaper robots are on the way, for example pneumatically powered robotic arms that can do a limited range of grasping and lifting operations and cost as little as £2000. This is well within the reach of small business. From Japan we already hear stories such as the one about a man who has installed three small injection moulding machines fed by robotic arms in his backyard shed.

Each morning he loads up pallets with plastics, programmes his robots, and switches on. Then he goes out to play golf for the day.

It is of course easy to be carried away by euphoria about these technologies without considering all their effects. The danger is to suppose that, because many industrial technologies are initially developed by large firms, there is only one possible outcome to this development process. As Schumacher reminded us in *Small is Beautiful*, there is always a choice in technology. Intermediate Technology Development Group, the organization he set up, has shown this time and again in their redesign of Western technologies to suit Third World users. In Britain, too, there is considerable rethinking about what types of technology are appropriate to industry, and how others can be adapted to make them more so. An example is the human-centred lathe. In the machine tool industry automation has aimed to replace the judgement and skill of the operator at all stages. Aiming to move away from this line of development, a group at the University of Manchester Institute of Science and Technology has designed an advanced 'human-centred machine tool' where sophisticated electronics take over the tedious or repetitive work, but the more creative stages are returned to the hands of the operator, assisted by the computer's brain where necessary.

Biotechnology

Another challenge to industry is the development of biotechnology, the industrial application of techniques for genetic manipulation of microbes, enzymes, plant and animal cultures. Biotechnology is already having an effect on the food, drink, fuel, pharmaceutical, and chemical industries. The early signs are that, with some exceptions, they are not techniques that will help small firms much. Biotechnology uses carefully engineered fermentation processes, in which there are some economies of scale. The techniques appear to be very expensive, employing small numbers of technicians but with no room for craftsmen. The exceptions are in the production of small quantities of specialized products, such as certain drugs, insect control systems, and biochemical engineering equipment.

For many small firms in these industries biotechnology will not make much difference. Many biotechnological industries are not new ones at all: the making of bread, beer, wine, cheese, yoghurt, the growing of mushrooms, and the composting of wastes all date back several thousand years. Smaller scale producers in these industries frequently use traditional processes to achieve a high quality control, the use of cheaper substitutes for raw materials, and speeded-up production, while regarding none of these as of primary importance.

The British cheese industry demonstrates this. Until the early twentieth century all cheeses were made in traditional round truckles in the farmhouse, matured in the cellar, then sold in local markets. Then two things changed, The Milk Marketing Board was formed, with producer control over all milk supplies and a dedication to factory production of cheese. Then the Second World War left cheese production under strict government control, and sacrificed quality for quantity. Before the war there were several thousand farmhouse cheese producers. After it there were only fifty, and many of these could more accurately be called small factories.

Not only has Britain lost most of its small scale producers, but it has also lost its rich variety of local cheeses. Dorset Blue Vinney and Scottish Dunlop may be gone forever. Cheddar now means any block of yellow factory produce. There is strong demand for good cheeses today, but many traditional cheesemaking skills have been lost. Consequently the last few years have brought a huge increase in imported cheeses from countries such as France which still take a pride in their traditional local cheeses, made by thousands of peasant producers. Meanwhile, instead of encouraging the start-up of small producers, the English Milk Marketing Board appears to stand in their way, in its self-appointed role of cheese market regulator, denying them milk at reasonable prices. Instead it teaches cheese sellers how to cut factory block cheese into segments that look as if they come from traditional farmhouse cheese.

Modern biotechnology has offered the cheese industry new forms of starter cultures, the cloning of rennet, the conversion of whey into new products, and the development of new coagulates. Cheesemaking can be speeded up and regulated, waste reduced, and substitute ingredients used. A tasty cheddar should mature for nine months in a cool cellar; the factory version can be 'ready' in a few weeks. These 'improvements' have been of little interest to the farmhouse producers except in so far as it reinforces their roles as makers of traditional high quality cheeses.

Changes in the business climate

The last decade has brought a number of other changes that affect the structure of industry in Britain.

Energy crisis

The first and most brutal of these was the OPEC crisis, in which the price of energy, a vital input in most industries, doubled in real terms between 1973 and 1979. Worst hit were the energy-intensive material-processing industries such as chemicals, ceramics, iron, and steel. However, it also gave several unexpected incentives to small local producers.

In the brick industry, for example, most of Britain's production, once the domain of small brickmakers, had passed into the hands of the London Brick Company and a few other large producers. Geared up for very large production, the London Brick Company installed huge Hoffmann kilns that allowed continuous firing of several million bricks at a time, consuming relatively little energy per brick. From the plants in the east Midlands the bricks were then freighted across the country. When the oil crisis struck the big plants found their transport costs for moving these bricks skyrocketed. Small plants had the advantage that they were mainly firing bricks from local clays for use on local building sites. Their transport costs being relatively low, the oil price rise hit them less.

In 1979 the second oil crisis struck and the economy went into recession, with the building industry suffering most. Now major brick manufacturers faced another problem. Their big kilns, which were energy efficient at high production levels, were very inflexible and extremely inefficient when used below capacity, raising unit costs of production. The small brickmakers did not use big continuous kilns, but relied on smaller intermittent kilns that could be fired as demand required. Consequently their energy efficiency did not fall and they were not left with large and costly stacks of bricks in their yards.

There is even some talk among craft brickmakers of returning to traditional methods, albeit using some improved techniques. As in the past, a temporary brickworks could be set up on a building site, the bricks formed from local clay and fired in traditional clamps (stacks of bricks with interlaid fuel) using improved pollution control techniques. This would mean more employment, low capital and energy costs, and no transport at all of the finished product.

Resource pressures

There is increasing concern for the pressures that are being put on scarce or non-renewable resources. More attention is being paid to ways of re-using or repairing products and recycling or recovering materials. Because these are basically service-type activities that reprocess small batches, use inventive handling techniques, and require opportunistic marketing, they are mainly the preserve of small firms. Repair and recycling in Britain was at its peak during the Second World War when everything that could be was mended and every bit of waste material, paper, fuel, or food, was carefully recycled. Since then the quantity of waste we generate has risen considerably and the recyclers have not kept up.

British recycling compares especially poorly with the record in North America and on the Continent. The scrap industry has a bad name for efficiency, honesty, and organization. Some of this is their own fault, some is due to the problems created by the packaging industry, the lack of interest shown by generators and buyers of processed scrap, the uninterest of the government in recycling, the low status of the trade itself, and the ignorance of the public.

Despite all these problems the trade is growing: the plastics recycling industry shows this. A wander around the less salubrious back streets of Birmingham may land you in front of an old warehouse with a heap of plastic rubbish outside. Inside a few people are sorting, clearing, and chopping up heaps of plastic wastes – polythene film, wrapping materials, misformed milk crates, old hose-pipe, and much more besides. This does not look like a factory of the future, and yet in many ways it contributes more to the country than do the high technology firms in the science park on the other side of town. It is increasing the material self-sufficiency of Britain, and at the same time makes a tidy profit for the owner.

The plant buys in plastic wastes, mainly from processors who make plastic items and packages. They sort out the main polymer types, and these are chopped into crumbs and made into pellets which are suitable feedstock for the plastics industry. There are many technical and marketing problems to solve before this becomes a major recycling industry, but when it does most recyclers will be tiny firms.

Newly industrialized countries

The 1970s and 1980s have also brought major changes in the international balance of industrial power, with the emergence of newly industrialized countries and other low cost producers. They are now dominating many of the industries that used once to be the preserve of older industrial countries such as Britain. The industries that have been hit hardest are shipbuilding, assembly work and textiles. Because they are labour intensive, this means big employment losses with effects right through the country. Some large British firms have reacted by closing plants at home and moving them to countries where labour is cheap.

Perhaps surprisingly, in many industries the survivors of this restructuring have been the smaller firms. This is because they have been more flexible, they offer higher value work which is not threatened by cheap imports, or else they are service companies – cleaners, repairers, caterers, transporters, wholesalers – and these services cannot be imported.

The wool fabric production of west Yorkshire is one industry that has been hit hard by cheaper imports of carpets, fabrics, and knitwear, mainly from the Mediterranean countries. One after another the big mills in the famous old wool towns have handed out redundancy notices and closed their doors. But in a picturesque valley of Pembrokeshire, Elwyn Davies is doing rather well with his small craft mill. Its old waterdrive wheel, its spinning frames, and its original hand looms are kept as a reminder of the past, and they present a sharp contrast to the modern high speed looms in use today. Mr Davies produces high-priced, carefully tailored clothing in Welsh wool, but there is nothing old-fashioned about his marketing methods, which are geared to the latest fashions and include sales in London and New York.

He is something of an exception. Who are the competitors that managed to drive all those hard-headed Yorkshiremen to the wall? In northern Italy is the region of Prato, an ancient centre of wool processing. Unlike Yorkshire, Prato has nearly 50,000 wool-workers in 10,000 tiny firms, most of them no more than garage-type workshops. But appearances can deceive. Inside these workshops is very modern small scale equipment, produced to specification by local machine-makers and capable of fast production of quality goods. Workshop managers are experts in technical production. In other aspects of management they are supported by a large number of banks in the area and by a network of merchants who obtain orders for work and divide them into suborders for each workshop. The system encourages rapid innovation, efficient machinery, and production of fabrics in response to new designs in a way many Yorkshire mill-owners would not have thought possible.

The infrastructure

Industry in Britain depends on the network of utilities there to support it. It is now becoming evident that after very little maintenance since the war much of Britain's infrastructure is in danger of decay or collapse. Inner city housing has been poorly maintained, railways, bridges, and some roads have been starved of money, and, out of sight underground, water mains and sewers are in a dreadful state. Much of this infrastructure was built in Victorian times by engineers who knew their jobs well and used good quality materials. Britain led the world in its rail network, its tunnels and bridges, and its underground assets. But no matter how good the original engineering, these structures cannot last forever without careful surveying and repairs. Since the war the country has been more concerned with new development on green-field sites, and not much money has gone into these old systems. Now many are near collapse and the sums needed to effect emergency repairs and longer term reconstruction are huge.

This is bad news for the taxpayer, but good news for the construction trade. This industry consists mainly of small firms using labour-intensive methods. There are roughly 5,000 collapses in roads in Britain each year, due to leaking sewers and pipes. Water scours the earth from around a pipe, making a bigger and bigger hole beneath the road, which is not noticed until one day a truck puts a wheel through the road surface. This means a lot of work for the local sewerage contractor with his truck and gang of workers, who do the unglamorous but important work of digging up and repairing the pipe, then patching up the road.

In the longer term major programmes are planned to survey sewers that have not yet collapsed, and where necessary to renovate them from the inside without ever digging up the road. This has been made possible by the development of some advanced remote-control technology and new, rather more glamorous, small firms are setting up for business in this area. There are firms which operate small closed-circuit television and video systems that can be pulled through sewers giving a picture of their structural condition as they go. Others provide special renovating materials which can be fed down a manhole to strengthen a weak sewer. Others operate 'sewer rats' – tiny remote-controlled robots that creep along sewers and carry out the needed pointing, welding, or cutting operations. There is a whole new industry in formation, and the innovative leaders are almost all small firms.

Changing attitudes in industry

So far in this chapter we have shown how new technologies and the changing business climate have brought with them new opportunities for small firms. At the same time there has been a switch in consumers' attitudes to what they want to buy, and in the attitudes of business people about how they want to operate.

The consumer revolt

Consumers, for so long neglected in Britain, are stirring at last. People are tiring of mass produced, standardized products which are also frequently bland, uninteresting, of low quality, and designed for a short life (except, paradoxically, in the case of foodstuffs which are designed for longer and longer lives). As British consumers become wealthier, and as they travel more and see how people live abroad, they are demanding better quality, special goods, regional products, and highly crafted work, for which they are prepared to pay more. This is good news for the small firms who cannot compete with mass production, but who specialize in quality work.

This has been especially noticeable in the food industry. When our ancestors moved to towns and cities after the industrial revolution they lost some of their links with the land and the food that it bears. The British rationing system during and after the Second World War contributed to an undiscriminating attitude to food. The large scale British food industry today prides itself on being efficient, but for high quality products it cannot compete with a country like France which has strong peasant traditions and many small scale producers.

But the future is brighter, as is already evident in the brewing industry. Once many tiny brewers made their own individual beers all over the country. These varied from lighter, sweeter northern ales to the heavily malted milds of the Black Country and the heavier stouts and porters of the south. As the bigger breweries began to assert themselves, they bought up many of the smaller plants and stopped brewing their local ales. This continued for some years, until by the 1970s nine large brewers accounted for 90% of the country's production and almost all the old brewing recipes had been lost. Furthermore, these big brewers changed the quality of the beer from a caskconditioned 'living' ale to a sweeter carbonated brew made with cheaper ingredients and served under pressure.

When some of the breweries decided to 'modernize' the pubs under their control the drinking public erupted in anger. The Campaign for Real Ale (CAMRA) was formed to oppose these changes, and it very quickly built up a strong national group of members lobbying for old-fashioned quality beers and traditional pubs to drink them in. As well as lobbying, CAMRA members like to drink and here was the basis of a market for small brewers of real ale.

Sure enough the new small brewers started setting up. In 1971 a small brewery in north Yorkshire commenced brewing again after twenty years. The next year two rural publicans built tiny plants to brew for their own customers. Then in 1974 a former Watney brewer set up a new plant to brew beer and set out to sell through free houses in the region. Since then small breweries have sprung up throughout the country, and there are now more than 150. They generally brew less than 10,000 barrels per year, which means that all together they are the equivalent of one big brewery.

Restructuring firms

Many business men and women in large firms across the country have been looking into the future, and they do not particularly like what they see there. They see that large scale producers can no longer claim a natural monopoly of production. The 'economies of scale' argument has been used too loosely in the past and needs rethinking. When brewing industry analysts claimed that efficient beer production meant plants with an output of a million barrels per annum, they did not make it clear that they had redefined the product itself to fit in with their large scale prescription. The small craft producers brew a fraction of this scale, but still make a tidy profit. In essence, large firms can produce some goods most efficiently while others are best made by small firms, and at last large firms' corporate planners are realizing this.

In many cases, through acquisition, merger, and aggressive buying, large firms find themselves today with a spread of activities that are not appropriate to their organization. This was easy to disguise during growth periods, but the recession has forced them to come to terms with the problems it can cause. The result has been a 'back to basics' approach in the most innovative firms, which has led to slimming down, laying off workers, axing peripheral activities, and buying in services instead of trying to provide them in-house. This has forced many large companies to change their attitude to small firms from outright antagonism or amused uninterest to real interest in them as potential subcontractors, services, or clients.

In the post-war years rubber components were produced by many small firms. During the 1960s and 1970s many were bought up by the two biggest rubber firms in Britain, competing to build up the biggest rubber conglomerate. Then came the oil crisis and leaner times. The corporate planners reviewed operations and decided to be ruthless: all activities outside the main stream would be discontinued and their staff laid off. Many employees were sacked, equipment put up for sale, and subsidiaries that had been bought up twenty years before were closed down.

What were these employees to do? Many bought up the machinery they had been using and set themselves up as small scale rubber fabricators, producing components in garages and small workshops. Often they can sell these components back to the original large companies, who are in a better position to market them. This restructured industry promises to be a healthy one, based on a better understanding of what big and small firms are each most efficient at.

After subsidiary divisions and peripheral activities have been shed by a large firm, its next step is to attack its own internal structure. Considerable effort is being spent in business schools today working out ways whereby entrepreneurial initiative and drive can be harnessed and used for the good of a large corporation. In the USA the best managed large firms have found that the most efficient way for them to manage their affairs is to create artificially small units within their large companies and encourage employees to be 'intrapreneurial', that is to act as entrepreneurs within an organization.

This usually involves setting up small, less organized teams for particular tasks, or groups working as independent profit centres; in its extreme case, many tiny firms work in a semi independent way under the marketing and financing umbrella of a larger organization. This is how many Japanese businesses have operated for centuries. The approach recognizes that there are actually economies of small scale in management, and aims to take advantage of these.

So economists have been forced to recognize that, as well as economies of large scale, there are economies of small scale. For example, small firms can motivate labour much more effectively and record fewer strikes and less down-time; management is much simpler and can often be done by a single person (though not all small firm founders are good managers); and small firms are more technically efficient at small batch production, jobbing work, producing high quality goods, and customizing services to homes, industry, and commerce. These are their natural roles. In filling them, however, many problems are encountered and these obstacles are the subject of Chapter 8.

We close this chapter as we began it, with the car industry. Mass assembly of cars can only be done on a huge scale in giant plants. But behind the facade of a few factories are thousands of medium and small firms who are just as much part of the car industry; they produce parts, manufacture subassemblies, provide a range of industrial services, customize cars, sell them, and repair them. Neither the large nor the small firms can operate efficiently without each other. The failures of the last decade have made it clear that a modern economy needs both.

CHAPTER 8
The rules of the game

How to play

'Monopoly' is one of the most popular board games. The players throw the dice to see how far each can move around the board. Whoever ends on a 'property' has the option to buy it, build on it, and charge the hapless opponents rent each time they land on it. Every time the players shake the dice, luck comes into it. But the dice is loaded; the easiest way to win at this game is to buy as many properties as possible, build as many dwellings, and wait for the money to come rolling in. The winner is the player who recognizes the way the rules are written and exploits them.

'Monopoly' is popular because it mirrors the economic game we play every day, whenever we earn money or spend it. A good game needs players of equal strength and a fair set of rules that does not favour one or another. In the last few chapters we have shown how small firms and local employment-generating technologies are staging a revival. Now let us consider the ways in which this revival is being obstructed by the rules of the game.

For the most part we are not asking how small firms and employment can be actively assisted, but rather how to remove the obstacles that prevent them helping themselves. We look in this chapter at four types of rules that can present obstacles: the ground rules that influence how firms operate; the way the country's business infrastructure helps or hinders them; the commercial institutions and how they affect business; and the ingrained attitudes of people. Finally we look at how the rules of the game could be rewritten.

More than a decade has passed since the two seminal works which sparked off public thinking about these issues. In 1972 Fritz Schumacher published a collection of essays under the title *Small is Beautiful* which came to be used as a catch-phrase of the movement. At about the same time the findings of the governmental Committee of Inquiry on Small Firms, usually known as the Bolton Report, was published. The Bolton Report made over sixty recommendations, mostly to government, but also to financiers, educators, large companies, and the media. Fritz Schumacher is dead, but he would doubtless be scathing about the lip service paid to the desirability of small firms without the rules being rewritten. John Bolton of the Bolton Report reviewed the situation in

1982 and noted that many changes had taken place over the decade, but much still needed to be done to catch up with other countries.

To mark the recent European Year of the Small Business, a report commissioned to describe the climate for small firms in each of the ten EEC countries found that Britain ranked second to last. The ranking measured how well the small firm fared when it came to employing labour, obtaining finance, investing, getting premises, paying taxes, and contributing to economic prosperity. The only country where there could have been a worse environment for small firms was Italy, but in that country there are many ways around discriminatory laws, taxes, and institutions that could not be recorded in the report. Quite clearly the rules of the game, despite all the books, reports, and Euro-commissions, still desperately need rewriting.

Ground rules for businesses

There are many operating rules for businesses, and a large number of them make it difficult for new small firms to generate employment and help revive British industry. We will do no more here than concentrate on a few main anachronisms.

One rule for labour

Why should the economy discourage unemployment? The levying of taxes on production is an important way to raise government revenues and an instrument of industrial policy. Alas, it also causes distortions. For example, financial companies, which have generally enjoyed high profits, pay only a quarter of the taxes paid by industrial companies, despite the fact that the latter have been very hard hit by the recession and that they generate far more employment in the country.

Within industry the burden of taxation falls heavily on labour compared with other sources of revenue. According to a report by the Institute for Fiscal Studies the proportion of revenue raised by direct labour taxes has doubled over the last 15 years. These taxes include National Insurance, private insurance, maternity, holiday, and other allowances. All of them are very important benefits, fundamental to our system of social democracy, but they should be viewed for what they are – a benefit to all society rather than a levy on a particular employer. If they continue to be charged as a payroll tax, then the employer will of course make every effort to replace workers with machinery.

Britain derives a higher proportion than most countries from income and social taxes, and a lower proportion from expenditure taxes.

Furthermore the income and social tax proportion has been increasing since 1979, despite the increase from 10% to 15% in the standard rate of VAT introduced in the 1979 Budget. In 1984 the total of taxes on income, social security, and profits amounted to 56% of the total tax revenue compared with 43% from expenditure taxes. Employers, concerned with the need to remain cost competitive, have to struggle against this increasing burden. In the more labour-intensive industries it can make a big difference between survival or failure.

Moreover, while British governments have assisted industry to restructure itself, little thought or money has been given to investment in the workforce. For example, when the wool textile industry was being restructured in the 1960s and 1970s, large amounts of money were made available to buy new looms and other equipment, but not a penny was available to retrain the labour force to operate them better. As a consequence large numbers of people were sacked from woollen mills, and their skills lost to the industry.

And another rule for capital

Why should the economy subsidize capital investment at the same time as taxing labour? Capital is subsidized because it is thought to be necessary to induce industrialists to update their equipment often. Unfortunately this can also have distorting effects on the way things are produced. As Samuel Brittan has written in the *Financial Times* (25 February 1982): 'At the same time that labour has been heavily taxed capital has been quite ridiculously subsidised.' The Institute for Fiscal Studies report found that, although capital charges and wages in industry are of similar magnitude, capital has been taxed only half as heavily as labour.

This is done in two ways. One is the tax code, which allows the cost of new equipment to be written off for tax purposes, usually over a year, although the equipment itself lasts for much longer. And if they are not making a taxable profit, manufacturers may sell their tax exemption to financing companies, then lease the equipment. Whatever the method, the real cost of new equipment is far less than the list price. The second distortion is via regional and industry restructuring schemes. Regional grants introduced since 1972 are related to capital investment, while the previous Regional Employment Premium system was tied to the number of jobs generated. In many industry schemes the government has offered money at concessionary rates to invest in equipment but very rarely to invest in the work-force. For example, from 1966 to 1978 over £100 million was spent on the machine tool industry, encouraging firms to re-equip and retool, this being seen as the key to a strong engineering sector. At the time this scheme was seen as a success, with a considerable amount of old machinery being scrapped and small firms closing down. And yet once again no money was allocated for retraining work-forces, whose qualifications and skill level have fallen far below that of their German competitors. Since those days the British machine tool industry has sadly declined in competitiveness, responding to the challenge of new technologies by continuing to produce equipment that is further and further out of date.

Blind faith in new and expensive technology, on the other hand, can lead to the sort of situation that exists in a major British car factory. The factory consists of two plants on the same site. The older one is cramped, noisy, and dirty, employing a large work-force in poor conditions. The newer plant relies on robots working with a much reduced staff. It is modern, clean, quiet, and well laid out. The robots, in which huge investments have been made, are given far better work conditions than their human equivalents.

Rules for new businesses

Why should the economy, stagnant and lacking initiative, want to discourage new business start-ups? Describing the economy as a forest in an earlier chapter, we compared big businesses to the established large trees and small businesses to the seedlings on the forest floor, the strongest of which must survive to ensure the plant stock of the future. When the government makes industrial policy, it is trying to manage the forest, improving the chances of one or other species for survival and growth. But rather than building nurseries for the seedlings, most government effort is devoted to propping up the rotting trunks of the failing forest giants.

Despite government rhetoric about new businesses and economic rebirth, precious little money goes to help start-ups, while huge sums are still doled out to traditional large scale industries. Ford's new plant in Wales attracted a government subsidy of £26,000 per job. The plans announced for the new Nissan motor plant in Sunderland indicate that £50 million of Japanese money could attract another £100 million in British grants. The justification for pumping in these huge sums is the widespread unemployment and poverty that would result from these industries' being allowed to fail. Rather than simply propping up failures, this assistance should also be designed to encourage ventures in new directions.

How is this to be done? Most struggling new businesses do not care too much for government incentives, because it usually takes too much work to qualify for them. However, what does worry them is the paperwork, regulations, laws, and restrictions concerning new business start-ups. A check-list of rules that a new business man or women needs to comply with runs to pages long, just at the time when he or she is likely to be concerned with business worries, selling difficulties, and cash problems. The VAT-men, planning officers, and health and safety inspectors who visit them are concerned only with the letter of the law and have little experience of business themselves. All these rules and regulations may well be necessary in a modern economy, but in the way they are applied they can form yet another barrier to entry to a business just when the economy is desperate for more growth. Small wonder, then, that this drives many otherwise law-abiding business starters underground into the black economy, where they are viewed as tax evaders, cowboys, or criminals.

But brighter signs are shining through the gloom of the forest, evidence that the government gardeners are at last learning how to build nurseries and to tend plants in them. There are now guidelines going out to government inspectors to be less bureaucratic and more helpful in dealing with new businesses. Certain government measures are designed specially to help start-ups, making finance easier to get. The most revolutionary and yet the cheapest of all has been the Enterprise Allowance, which gives would-be business starters who are out of work the equivalent of their unemployment benefit for a year as they set the venture in motion. This is likely to result in a large number of low-capital businesses: many will fail, but a number will survive to ensure the future of the forest.

Another bright sign is the growth of voluntary support services. There are today nearly 200 local enterprise trusts in most regions of Britain, and they deal with over a quarter of a million of enquiries about new businesses each year. Small firms advisers, whether they be enterprise trust officers, local government industrial development officers, bank managers, accountants, or others, are now numbered in the thousands. Sometimes they work alone, in other cases nursery work spaces have been set up, combining all the technical and business skills under one roof. The story of these support services is told in Chapter 9.

And rules for small firms

Why should an economy wish to penalize small firms and encourage large ones? This is what happens: those new business men and women who survive their first year or two, despite all the barriers to entry and the natural commercial risk, will still find the rules of the game written against them.

Government policy moves very slowly, especially in those areas like technology, industry and employment where it is important that things happen fast. As a legacy of the post-war era when it was thought that you had to be big to be efficient, government assistance still bypasses smaller firms and legislation discriminates against them. This was a major complaint of the Bolton Report on small firms in 1971.

There have been some improvements since then, but all too often policy assumes no particular role for small firms, aid bypasses them, and legislation obstructs them. Government spending goes mainly to those industries like iron and steel, motor vehicles, and aerospace that are dominated by large firms. One complaint of the ailing iron foundry trade was how much the small firms that comprise the foundation of the industry had to spend to meet health and safety regulations. Their argument was not that there should be no regulations, but rather that those in force had been designed for larger scale continuous foundry operations and were mostly inappropriate to the smaller batch producers. In earlier chapters we have put the argument for appropriate technologies; here is a similar argument for appropriate government policies.

One reason why such inappropriate policies persist in Britain is the way that industry makes its needs felt. The groups best at lobbying and with the greatest influence over government are the powerful industrial associations such as the Confederation of British Industry and the Trades Union Congress, representing one face or other of large organization interests. Without the sophisticated lobbying system of the USA or the small party coalitions of the Continent, the British system of government seems unable to reflect the needs of small producers. Diverse and decentralized, small firms here present no unified face and are therefore too often overlooked by policymakers.

The business infrastructure

'Go to jail ... do not collect £200', says one of the cards in the 'Monopoly' set. Whether a player is unlucky enough to pick up this card depends on the fall of the dice and on the way the board is designed. In this section we look at some underlying aspects of the country's business infrastructure that determine where its players can move.

Where businesses locate

Britain has a regional policy, a system of administration, and a transport network that encourages a business to locate very centrally, to the detriment of certain regions.

One of the underlying business myths of post-war years that arose with the 'economies of scale' argument, has been that it is best to concentrate operations in a large central plant, preferably on a green-field site near a motorway. From there goods are trucked long distances to regional warehouses and eventually on to retailers. This can lead to ridiculous siting and transporting decisions being made in the name of rational business management. Schumacher gave a favourite example in his film *The Other Way*. Filming from a roundabout on the A1, he showed a lorry full of biscuits on its way from London headed north. Then, panning the camera around to the other lane, he caught another lorry, with Scottish markings, also carrying biscuits, headed south. The only explanation of this, he concluded wryly, was that there must be something intrinsic about biscuits that necessitated they be taken on a 500-mile ride before being fit for eating.

This sort of situation has been encouraged by a regional policy that induces large imported turnkey plants in regions designated as deserving of largesse from Whitehall, without particular concern as to where the proposed markets are, the social costs of transport incurred, and the contribution to local development. Thus it is decided that cars shall be made in Sutherland, aluminium in Invergordon, and video recorders in South Wales, rather than promoting in all three regions smaller businesses that could, for example, service cars, recycle aluminium, and repair electronic goods.

Strategically, too, this is a curious policy. If Hitler had his time again, he would have to bomb only a small number of vital factories to put Britain out of the war altogether.

Over the years we have moved from low energy transport networks for moving freight (canals, then railways) to high energy systems (motorways and aircraft). The way that roads are paid for consistently undervalues the real cost of moving freight around the country. Given the low transport charges they pay, the London and Scottish biscuit factories may be quite rational to freight their biscuits the distance they do. They do not have to pay the full price: the noise and air pollution from their lorries, the dangers of accident or death, the damage to roads and buildings from their vibrations, the despoiling of countryside criss-crossed by roads, and the frayed nerves of the people who live along the way.

This system is a hard one to change, not just because of the road lobby but also because increasing freight charges to reflect these social costs would make British production more expensive and less competitive. So changes will have to be carefully thought out, and staggered over a long period to let a new pattern of decentralized small industry establish itself, feeding local markets and providing local services where appropriate.

Who owns businesses

As in the game of 'Monopoly', the game of business encourages the concentration of ownership into the hands of a few. This has happened at the expense of competition, consumer choice, and efficiency in the economy. One reason has been the lax attitude of the government to competition, and another their attempts at industry restructuring.

The growth of firms since the war has been partly accomplished by business takeovers and mergers, successively concentrating the ownership of firms in fewer and fewer hands. Most of these takeovers were intended to obliterate competitors and they have left many industries with only one or two major producers, who are in a position to make unilateral choices about product range, design, and price that will affect us all. Much of this would never have been allowed in the USA, which has strong anti-trust legislation. But in Britain the powers of the Office of Fair Trading and the Monopolies and Mergers Commission, subject as they are to political direction, are no match for the industrial giants.

In the 1960s the government set up the Industrial Reorganisation Commission to help 'rationalize' certain industries by inducing the closure of smaller units. The Commission was noticeably unsuccessful in its attempts to make industry more efficient and has since been disbanded, but it did have the effect of concentrating ownership in certain industries. Another reason for this concentration has been the evolving British financial system, which is itself highly centralized and encourages other businesses to set up on national lines, with head-quarters in the City of London.

Time has shown that few of the firms involved in the mergers and takeovers of the 1960s and 1970s did particularly well out of them, most actually suffering falling profits because of these activities. They tended to create bureaucratic monsters out of vigorous young firms, losing much of the entrepreneurial initiative that the latter breed. Considerable effort is now being devoted to finding ways whereby this initiative can be recaptured and harnessed to the large corporation by means of intrapreneurial organization, productivity circles, profit centres, and dividend sharing agreements. This basically means trying to recreate the organization of the small firm within the ownership structure of the large.

Quite apart from the question of efficiency, concentrated ownership means that fewer and fewer people are making the strategic decisions which affect others, something that is not healthy in our society. A small number of holding companies control more and more of the economy, while fewer and fewer people control these holding companies. A true political democracy requires full economic participation, and the reins of industry in the hands of a few does not help this.

What industry uses and what it wastes

Perhaps the most crucial ground rules of all concern how the country's resources are used. A highly populated island nation such as Britain can scarcely afford to encourage misuse of primary resources like coal, wood, ores, and gas. Neither can we overlook those modern strategic materials like certain precious metals that we need but cannot produce locally. Unfortunately the misnamed Department of the Environment ignores these issues, and we still do not have a resources and recycling policy.

As a consequence, industry receives no encouragement to avoid making new glass bottles, producing new plastic feedstocks, rolling new paper, and weaving new fibres, rather than recycling the old. What happens to the old materials, known as 'rubbish' or 'waste' in Britain but considered an important raw material overseas? We send waste fabrics to Italy, scrap steel to Spain, plastic rubbish to Holland, and old machinery to Africa. Britain is probably the biggest exporter of scrap materials in the world, and has one of the worst recycling records.

The region that once led the world in textile recycling was Morley in west Yorkshire. Here were woven from recycled fibre the army blankets, donkey jackets, the duffel coats that kept generations warm as they worked or slept. Today the mills of Morley are abandoned and silent, the workers sacked, the machinery sold, the skills forgotten. Instead we sell our waste textiles to Italy, where there is a vigorous and imaginative recycling industry. Back they come as imports, Italian woven coats and Italian knitted jerseys, bright and fashionable clothing made from materials we can no longer use.

Recycling is a difficult process, increasingly so as new materials and new packaging become more common. Once the preserve of the gypsy and the rag-and-bone man, recycling is now becoming the trade of the chemist and the engineer, as new ways to collect, separate, and purify materials are found. There is a role for the government to stimulate research here, encouraging new thinking and upgrading the image of the recycler. Official standards need to be set for the quality of virgin and recycled materials used as feedstocks, and recyclers can be helped to reach these standards.

Help is particularly needed to simplify the recycling of packaging: identifying the different materials used (especially important for plastics, where there are dozens of different polymer types), avoiding mixing incompatible materials, and avoiding the sticky labels, metal caps, and other bits and pieces that make life hell for the recycler. The packaging industry has been given years to do this itself and, sadly, has failed.

With this sort of assistance the proportion of waste materials that can be recycled will rise from today's low levels (3% of plastics, 3% of glass, 9% of lead, 24% of aluminium, 29% of copper). It will be impractical to return all the materials to their original uses, and here new uses may be found. The glass industry showed this was possible when it commissioned Cardiff University to develop uses for pulverized waste glass. They came up with a range of interesting new products, including attractive tiles made of polished glass fragments. Sadly, nothing more has come of this venture.

Just as the recycling industry needs help, so too does the repair industry. Too many appliances built or imported into the UK today are designed to have only a short life, and when they break down are not meant to be repaired. Irons have to be thrown away because of a burnt-out element. Washing machines are scrapped because the timer breaks down. A Japanese television set selling in Britain is designed to short circuit when a repair man tries to trace a fault. There are a few exceptions to this trend – quality appliances where durability is guaranteed. But most of the electrical goods industry urgently needs to be shown standards for both durability and repair-ability.

The commercial institutions and their operations

The safest player on the 'Monopoly' board is the banker. The rules the banker operates under assure this person of a steady income at the other players' expense. The only institutions in 'Monopoly' are the jail and the bank, but real life is more complex. In this section we examine the way our commercial institutions set their operating rules, and how these too often act as obstacles for small firms, appropriate technologies, and employment generation.

Buying and selling

The small enterprise is often discriminated against when it comes to buying and selling. Small firms almost invariably have to pay far more for their raw materials and services than do large firms, who as bulk buyers can count on enormous discounts. Sometimes this merely reflects the lower cost in packing and shipping a large order; often, though, it also reflects the fact that a supplier will do anything to keep a large customer without worrying about a smaller one. Small firms complain about this but not much has been done to improve matters.

When it comes to selling, small suppliers may again find it difficult to get reasonable prices for their work. In the clothing industry, for example, it is unrealistic for most small manufacturers to sell direct to the public; most of them supply the large retail stores which sell most of the country's clothes. Given the structure of a few large chains there is not much choice about who to sell to. The purchasing offices of these large shops are very powerful: whether they like or dislike a new garment can mean prosperity or failure for a new business venture. Not surprisingly, small business men and women are full of complaints about these buyers. Some as a matter of policy will buy only from large, well-established suppliers. Others, rather rarer, are more innovative in what they will consider.

Marks and Spencer have an innovative attitude. If convinced by the potential of a product they will undertake to help the supplier produce it, advising on quality control, production management, use of raw materials, packaging, and so on. They have a team of 300 technologists who act as industrial advisers. If a supplier follows their instructions, long order runs are assured. This positive, if paternal, leadership role is rather characteristic of older Japanese trading houses and their relations with clusters of small suppliers. Even today over half of Japanese exports come from small firms, with the big merchandizing organizations providing finance and marketing capabilities.

In Britain the large officeware manufacturer Pitney Bowes, with its wide network of sales people and outlets, is trying a new 'piggyback' distribution system whereby it will remarket appropriate products made by small firms. Another method is the market co-operative. Many tiny Italian producers, especially in traditional craft sectors like furniture, textiles, or machine tools, form co-operatives to market their wares jointly, both within the country and overseas.

The biggest single customer in Britain is the government, spending over £20 billion annually. Governments usually purchase by tender, but these can involve considerable trouble for the supplier and are most suited to large firms which have professional tendering officers. For example, one tender document for occasional tables worth £5,000, intended for junior management in the Foreign Office, ran to 29 pages in length and referred the reader to 22 other documents. Most Western governments, aware that this system favours large firms, have set up systems to ensure smaller ones get their share. In the USA a fixed proportion of certain contracts must go to small firms. Britain stands almost alone in having no buying guidelines: recent government purchasing reviews have confirmed this problem, but done nothing to address it.

The financial sector

Why is Britain so 'advanced' in its financial sector, and yet slipping so badly behind in its industry? The country has a system of banks, insurance companies, brokers, and other institutions unparalleled in the world. Yet small firms still complain they cannot get finance. One problem is that the City of London is too international in its outlook for the good of Britain. So efficient is the system that it can invest British funds in profitable uses throughout the world. This can mean high returns, but it does not ensure finance for our own troubled regions and industries.

The more recent development of unit trusts and pension funds has exacerbated this. Huge sums of money are mobilized from employees' contributions and small savers. Where they are invested depends on the judgement of a select band of fund managers, who are now so powerful that they can interfere in industrial decisions. Between 1975 and 1981 pension funds and insurance companies increased their ownership of publicly quoted shares from one-third to one-half.

Their investment power breeds problems. John Bolton of the Bolton Report on Small Firms estimated that in 1970 only 0.3% of their funds went to small firms, and today it is still a tiny fraction. They can also leach money out of a poor community. For example, the retired miner living in the blighted Rhondda Valley has his savings invested in the NUM pension funds. In 1981 this fund had assets of £650 million, it was extremely powerful – and it re-invested only £12 million in property in Wales. In this it is typical of most large funds.

Some changes are taking place. Banks and trusts are becoming more aware of their responsibility in these matters. There have been a number of attempts at setting up community banks, local trust funds, and local authority enterprise boards. However, we have still a long way to go before we reach the situation in the USA, where each year businesses give or lend millions of dollars to community funds, or Germany, where there are 3,000 local savings and co-operative banks, raising funds and making investments in their localities.

Science, education, and training

The institutions that determine the way we think about work and money can either be a focus of inertia in the economy or else powerful forces for change. There are two ways in which these bodies influence production. One is via the application of money and brain-power to the development of processes and products. This question of research and development is discussed in Chapter 5.

The second way is more direct. Just as research and development should lead to better technologies, so education and training (among its other functions) should lead to a better work-force. Unfortunately, British educational establishments at all levels have been sadly deficient in this. Many schools provide little instruction about industrial and commercial systems, and nothing about how businesses are operated. This has always been a failing, but the rise of youth unemployment has made it a crucial one; young school-leavers can pass directly from school to dole queue without any realistic work experience. The long-term implications of a generation who have never held jobs are horrifying. It means a generation who do not know how the system works, who may never participate in it, who may never benefit from it, and who have no stake in it.

The recession has exposed a dangerously thin programme of technical training, and the apprenticeship system is now in tatters. In most industries it is difficult for the young to get jobs that carry any formal technical and vocational training. If these industries survive, there will be dangerous shortages of British skills in the future, and a large unskilled labour force without jobs. Already we are in the unenviable situation where half our workforce has no educational or vocational training at all, twice as many as in Germany or the USA.

At tertiary level, technical training has always had low status in Britain, and those polytechnics and universities that are best at it have been hardest hit by cuts in government support.

Technical universities like Salford and Aston have responded by selling more of their services to industry and are in a buoyant state. However, they have had to suffer the ignominy of knowing that they rank the lowest in the government's list of priorities.

With a few exceptions, such as Imperial College, London, universities have played a minor role in economic development in Britain. Their relationships with the commercial sector are unsophisticated, and their research and teaching priorities do not always accord with the country's needs. Only 16% of university graduates go into business at all, and only a fraction of these into small firms. Technical degrees such as engineering confer relatively low status, in contrast with countries such as Germany, where engineers are important people and where many more university graduates are to be found in businesses of all sizes.

The business culture

Why are the rules of the game stacked against a small firm revival? Why does government industrial policy continue to be so reactionary? Why are educational, financial, and commercial institutions allowed to erect barriers against new initiatives? When the rule-making is scrutinized the only reason these barriers persist is that we, the general public, accept them.

Ten years after the Bolton Report on Small Firms was published, John Bolton talked about some of the Commission's experiences:

> Perhaps the most important and alarming realization that emerged during the course of our inquiry was the sublime state of indifference, in the United Kingdom generally, to so vital a sector of the economy. As we had gone around the world studying attitudes and developments in other countries, the Japanese could not believe there was no specialist department in Whitehall to monitor the health of the small firm sector; the French could not understand why a British M. Poujadt; had not emerged to give small business some real political clout; the Germans were surprised that Britain did not have a network of training facilities geared to small enterprise; and the Americans were astonished by Britain's failure ... to place small firms, as they do, on the same emotional plane as "motherhood and the flag".

This indifference is seen at all levels in Britain. A study of issues raised in Westminster during the 1970s showed that there were as many parliamentary questions about dogs as there were about small business. But it is unfair to criticize the politicians, the civil servants, the industrialists, the technicians, the teachers, and the workers, as we have done throughout this book, without pointing out that they can do no more than reflect the attitudes of the great British public.

There is a deep-rooted feeling running through society that, although things are bad, change is probably not desirable, and even if it were desirable it is probably not possible. People have gut feelings, based on the past and their own experiences. If analysed, these gut feelings say: big is always efficient, unemployment is inevitable, technology is pre-ordained, industry cannot be revived, and commerce is dirty.

This British attitude is very different from that found in many successful countries today. Why should this be so? Some analysts argue that it is a hangover from the days of the industrial revolution when new commercial fortunes were made, but the newly rich tried to ape the aristocracy of the day instead of founding a new business culture as in the USA. Their money safely in the bank, these businessmen educated their sons not in engineering but in classics, and, instead of putting them to a career in the factories and mills, tried to set them up as genteel landowners. The outcome is evident today.

Changing the rules

Can the system ever really be changed to remove the obstacles we have identified and encourage an industrial revival? Can the rules of the game be rewritten?

The easy answer, of course, is 'yes'. In time anything can be changed. In a decade the views of today's civil servants and politicians, industrialists and workers, educators and pupils, may seem as absurd as do many Victorian views today. The vanguard of change is already evident in the wave of new initiatives sweeping across the country that we turn to in Chapter 9.

What is a realistic timetable in which to achieve the changes we have talked about? For the most part the operating rules of the economy could be changed tomorrow. In practice it will take at least an election or two to achieve this. But rules and regulations need no more than a change on the statute book, a different operating procedure, or new directions to staff. The 1990s do not have to be bound by the rules of the 1980s.

Changing the country's infrastructure and its institutions is much slower, because it involves both physical reconstruction and the rebuilding of organizations. This does not happen quickly, nor should it, because mistakes in this area can be very costly, as we have seen in the past. It will be the twenty-first century before we can expect major improvements here.

But what could take longest of all to change, and what is most important of all, are the deeply felt attitudes of the population. Most people learn only from their own experiences and from what their parents and teachers pass on to them. A complete change might take a generation. By then it will be a whole new game.

CHAPTER 9

Taking the initiative

Breakdown of community solidarity

The beginning of mass transport was in Britain. As far back as the third decade of the last century George Stephenson was linking centres of population and industry by means of railways. Between Whitby and Pickering in North Yorkshire, for example, a stage coach link had operated only since 1795, when in 1832 Stephenson was invited by a group of Whitby townsmen to survey a route and advise on a possible railway. They met in a local pub and at the end of the evening sufficient cash was collected to pay Stephenson. In his report proposing construction of a twenty-four-mile route at a cost of £80,000, he wrote of the traffic that he expected to develop. He then continued, 'It appears to me to be deserving of the most cordial support of all parties, of the inhabitants of Whitby, from its ensuring both to the town and harbour an increased activity of trade; of the landowners, from its affording them the opportunity of converting an immense barren tract into fertile land; of the inhabitants of Pickering and all the towns and villages of the Northern District of Yorkshire, from its enabling them to obtain coal as well as other commodities imported into Whitby, at a considerably lower price; and lastly, from the prospect (amounting, I conceive, to a certainty) that the revenue will amply remunerate the proprietors for the money invested.' When the report was discussed at a public meeting in the Angel Inn, Whitby, a share list was opened and nearly £30,000 was subscribed by local residents on the spot: less than two years later the railway was in regular operation.

   This is a fairly typical example of the way that community solidarity and co-operation played a critically important role in the development of industrial society in Britain a century and a half ago. It operated in a different context with the formation of trade unions. 'They helped everyone his neighbour, and everyone said to his brother be of good courage', is a quotation from a union rule book of the same period which expresses the spirit of brotherhood that inspired the early unions. Frequently that spirit was demonstrated in warm fellowship amongst workers as they gathered in a pub room to pay their weekly dues. Nowadays many members' subscriptions are deducted from their pay and paid en bloc by the company to the union. A financial transaction has taken over from a human relationship between union members and branch officers. Today investment – and disinvestment – in railways, and a great deal of industry and commerce, is either in the hands of government agencies or in massive faceless financial institutions. Again, a disembodied financial transaction has displaced co-operation between individuals who elected to share both risk and opportunity. Union members and investors were motivated by self-interest, but it was set in a framework of comradeship and loyalty to the common good.

This process of dehumanization has not been confined to union membership and investment. It has spread widely through much of the life of local communities. This is frequently recognized by people who compare today with, say, the thirties. David Durston has written (in a private communication) about such people in the Black Country. They admit, he says, how cramped and small the houses were, how lacking in amenities, how poor people were, how hard life was; but they still insist that things have deteriorated. This is because they feel deeply the loss of the sense of community solidarity that supported them in the face of hardship and grinding poverty, and provided them with most of their satisfactions in life. The loss is human and priceless. As David Durston goes on to say, the attractiveness and glamour of the things that the industrial system has produced makes it hard for those who have not experienced this sense of loss to take community spirit, solidarity, and co-operation seriously. Frequently when it is mentioned there are comments such as 'romantic nonsense'.

In a report drawn up for the European Commission entitled *An Energy Efficient Society*, Jean Saint-Geours and a group of independent experts concluded that Europe can meet the challenge of developing an energy-efficient society without putting at risk the values, traditions, well-being, and liberties that Europeans hold dear. However, they make it quite clear that such a development cannot be directed by an economic planning office. It demands a high degree of individual initiative and responsibility; and a long-term energy-saving policy can only succeed if it has large public support. The same conclusion would apply equally to a society in which efficient small scale production and service activity is to reverse the trend towards bigger units and centralized control. Previous chapters have considered technical, economic, and legislative factors ('rules of the game') that are involved in changing this direction of economic development. This chapter is concerned with the socio-environmental problem of gaining widespread public support in a society which has lost much of its community solidarity and co-operative spirit except when confronted by disaster. It is certain that we cannot resurrect the past in its old form. It was the product of its own time. If we are to establish new community values that alone can satisfy the need for local solidarity, we need a wide range of initiatives which demonstrate that, under the conditions of the late twentieth century, a spirit is still alive which can be released to good effect.

New hope for energy

In the past few years there has been a flowering of a wide variety of experimental local initiatives which have drawn, in the main, on the enthusiasm and skills of local people to mobilize material and financial resources within their locality. Most have encountered great difficulty in becoming established, and by no means all have survived in an environment that is in many different respects antipathetic to unconventional alternatives to received economic, social, and political wisdom. However, enough have survived to encourage great hope that sufficient co-operation and enthusiasm can be generated for a rich, fresh growth of community-based enterprises of many different kinds.

A remarkable example was started in Newport and Nevern in South Wales in 1979 by Dr Brian John, a geographer and local resident. The total population of these two communities is only about 1,500. Nevertheless, with the leadership of Dr John and a few enthusiastic friends, they were able within two years to build up a membership of about 100 in the local energy group dedicated to the task of greatly increasing their own energy efficiency and that of their community. One stimulus was their estimate that the community's total expenditure on domestic energy alone was £114 million each year, all of which was being paid to suppliers outside their area. Another was the realization that they were spending money on non-renewable fossil fuels and wasting half of all they bought through inefficient use. It seemed both socially responsible and economic common sense to invest in things like draught excluders and insulation in order to reduce consumption of fuel, thereby retaining something like £100,000 per annum – nearly £200 per household – to use for worthwhile purposes in their community. Within about two years they had made such good progress on their housing stock that they were able to turn their attention to public buildings, such as the local school.

This pioneering effort in a small Welsh rural community has been followed in other British towns and areas. When a group took up the idea in Oxford they estimated that there was a potential saving of about £10 million that would be liberated for other uses by the population. They also recognized that the capital investment in fuel saving would result in substantial increased employment in Oxford and elsewhere, and that there would be a knock-on employment benefit when the £10 million saved was released for other purposes. Although there has been a modest uptake of this type of local initiative in Britain it has been considerably less impressive than has occurred in many parts of the United States.

Domestic energy offers only about half of the total potential for savings. Our industries also are wasteful users. Some unnecessary energy use in industries results from the low proportion of domestic refuse that is recycled. Local energy groups are able to tackle that problem by organizing recycling projects for glass, tinplate, aluminium, paper, and plastics. This all helps to halve national energy consumption, which cost about £25 billion in 1984 and is equivalent to more than 4 tonnes per annum for each man, woman, and child. We need to 'Halve it!' in the next few decades if we are to avoid becoming impoverished by our wastefulness, as fossil energy becomes scarce and prices ration it in favour of the rich.

Although in this book we are primarily concerned with local initiatives which are material and resource conserving, and which demonstrate the feasibility of a wide variety of employment and wealth creating activities being carried out on a small scale, it would be a mistake to overlook the many community activities which can be broadly described as 'social'. Whether it be groups of parents collaborating in setting up alternative schools for local children because they are dissatisfied with what the county authority provides, or a group that has formed a self-build housing association, such co-operative ventures give credence to the conviction that there are still viable options in most parts of life from which to choose. One very important lesson that all such initiatives demonstrate is that we are in reality much less dependent upon money in dealing with our needs than we are generally led to believe in a money dominated society. The parents do not have to afford exorbitant school fees; they share their time and talents instead. The high cost of buying a decent house was out of reach of the self-builders; but by mutual co-operation they were able to achieve what otherwise was impossible.

New hope for homes

A rapidly growing self-build movement has developed in the past ten years. In November 1974 Mike Lundberg advertised in his local Swindon newspaper for people willing to build their own houses. Sixteen people replied. Most lived in council houses, some in caravans, and two owned their own houses, which were for one reason or other unsatisfactory. They met many frustrations before finding a site and gaining approval for a small development: not until the autumn of 1976 were they able to start building. All twelve houses, built by the Downsview Self Build Housing Association on a site purchased from the local authority, were completed in 1978 to the great satisfaction of the members. Building finance was provided by the local authority; and running expenses and hire of equipment was paid for out of a loan and the members' modest weekly subscriptions. The loan was paid off by mortgages provided by various building societies to individual householders on completion. The group registered under the Industrial and Provident Societies Act (1965) and joined the National Federation of Housing Associations. Its operation was in accordance with the Model Rules of the Act and the working regulations of the NFHA self-build manual. Members were organized for specific tasks and duties. Strict rules, with fines levied against offenders, were applied to the weekend and summer evening working hours. Not only did a dozen families obtain greatly improved housing at a fraction of the normal market price, but they proved to themselves and to many other people that apparent impossibilities can be achieved when people act together with resolution.

Even though the work done by self-builders is unpaid in a conventional sense, the wealth created is in no sense unreal, being reflected in the resale value of the property. Neither was the education-wealth created by parent-teachers any less desirable simply because no money exchanges took place between parents and teachers. Part of the need in rebuilding community self-confidence is the denial of dependence on increasing amounts of money as the sole means of progress.

There is nowhere more important for this lesson to be learned than in affluent societies. In Beaconsfield, Buckinghamshire, some members of local churches were concerned at the lack of evening meeting places for young people and by the inadequate daytime provision of cafe facilities in New Beaconsfield. High property values and rents, and high wage rates, made a conventional business 'uneconomic'. Both needs were, however, met when the Beaconsfield Coffee House was set up and operated with a minimum amount of paid staff and a great deal of volunteer help. After a few years of operation all loans were repaid and a second expansion phase was undertaken.

New hope for rural areas

The depopulation of rural Britain and its villages has resulted in many traditional local activities becoming 'uneconomic' in conventional money terms. Schools, shops, filling stations, post offices, transport services – even village churches – have been affected and many have been closed down. The story of Llanaclhaearn in the Llyn Peninsula of North Wales is worth telling. In 1970 the local education authority decided to close the village school with its 27 pupils. The bus service to

Nefyn ceased. Two butcher's shops in the area closed down and three months later the last quarry ceased operation. Between 1961 and 1971 the population of the area fell from 1,250 to 1,050 as people moved away to find employment elsewhere. Aelhaearn itself has a population of only 300. With the leadership of the local doctor, Carl Clowes, a community association – Antur Aelhaearn – was formed in 1972 following a successful fight to prevent the school closure. This was the first community co-operative in Britain formed to revitalize its area. A site was acquired and a building erected to provide work-space in the village. A knitwear business was started which by 1979 had a turnover of £50,000 and employed five full-time staff and thirty part-time outworkers. Members of the electoral roll are entitled to one share of £1 each, ensuring that control of Antur is spread throughout the village. They meet annually and elect a ruling body of twelve. The greater part of capital is raised as loans in units of £5, but some money is also raised through fetes and other social events. All is directed towards improvement of the life and prospects of the village and surrounding area.

There is more to life than the essentials of housing, food, and clothing. Everyone needs recreation; and while the industrialization of leisure may have reduced some types of community activity, new opportunities have been created. Some have been very remarkable demonstrations of enthusiastic co-operation in work without monetary reward.

The revival of Stephenson's Whitby to Pickering Railway as a preservation project – the North Yorkshire Moors Railway – is not untypical of the mushrooming of railway preservation societies throughout Britain. In 1967, two years after closure of the line by Dr Beeching, a well-supported public meeting decided to restore services. Imaginative fund raising efforts, combined with the commitment of many enthusiasts to contribute 'sweat capital' helped persuade the North Riding County Council and the English Tourist Board to lend support to the purchase of the eighteen miles of track and other physical assets. Steam and diesel locomotives, with other rolling stock in various states of disrepair, were gathered from many parts of Britain. After heroic work by volunteers, repairing, reconditioning, and rebuilding track, buildings, locomotives, carriages, wagons, and signalling equipment, passenger services were resumed in May 1973 with a formal opening by the Duchess of Kent. Passing through the heart of the beautiful North Yorkshire Moors National Park, the NYMR is one of the largest and most popular tourist attractions in the North of England. With the passage of time there has been a remarkable development of prosperity and different forms of employment in the area. It is largely a result of this railway revival, which was made possible through the enthusiasm and hard work of many men and women who find recreation and satisfaction in running a railway; and also from the 120,000 passengers each year that enjoy the experience of riding in an old steam train through Newtondale.

New hope for inner cities

Lest an impression be given that community initiatives occur only in rural areas, some examples must be mentioned of major city projects. There are a great many to choose from in different parts of Britain. Towards the end of 1977 the Glasgow Council for Voluntary Services began what they have referred to as 'the best scrounging campaign ever mounted'. The concept of 'Goodwill Incorporated, Glasgow' was inspired by Goodwill Industries in the USA which traded in restored and reconditioned secondhand goods in department-style stores. The first step taken by the GCVS was to obtain a secondee from IBM, who identified a gap between expensive household products sold by high street traders on the one hand and 'charity' shops on the other. With assistance from the Regional Council and the Scottish Development Agency a building was obtained and renovated providing a shop front and a reconditioning workshop at the rear. Official finance, however, left a gap to be filled from private sources and the massive 'scrounging campaign' brought in all kinds of valuable material and equipment. The shop in operation is bright and attractive to look at, and is completely in keeping with its location in a busy shopping area of the city. Vacant premises, surplus materials, and equipment were given a new life, which, when combined with human energy, skill, and enthusiasm, provided a low cost source of supply of good quality household goods that were surplus to their original owners' requirements.

A different kind of urban initiative was set up following the closure of Courtauld's big factory in Skelmersdale, Merseyside, in 1976. It is called the Association of North West Worker Industries and its aim is the creation of a number of small scale worker co-operatives. The first three co-ops were a company employing 28 redundant people to repair and refurbish school furniture for the Lancashire County Council; a small metal fabricating business employing 15 people; and a furniture manufacturing firm which had gone bankrupt and was bought out by its workers. Although each is a separate private registered company, operating as co-ops they are linked in a two-tier arrangement with a 'holding company' which provides overall financial control and expertise, and the capacity for initiating new schemes. Hawkesbury Investments Ltd, the holding company, is registered as a charitable trust which supplements and reinforces the limited business skills and acumen of the individual co-op members.

Hyson Green, Nottingham, was built in the 1960s, an ugly concrete estate of flats and maisonettes. Garages underneath the accommodation soon became vandalized and subject to many forms of misuse. An approach by the tenants' association to the city council in which they proposed conversion of the garages to small starter workshops was turned down. The council's proposal to convert them to an indoor bowling alley of international standards was withdrawn in the face of public opposition and spiralling costs. In May 1979 it was mutually agreed by the tenants and the council that the garages should be converted for the benefit of the estate community – a portion to be used for new small businesses set up by tenants who had completed a new enterprise training programme; a second part to be a training workshop for 55 trainees; and a third portion to be let for established commercial businesses, with rents providing revenue for the overall scheme. A limited company is fully responsible for the project: 74 shares are held by the tenants' association and 266 by the city council, the estate landlords. The board of directors numbers seven: three elected by the tenants, two by the council, and two 'expert' directors elected by the tenant and council directors.

Some idea of the range of local economic initiatives that can be formed now in many major cities can be obtained in Bristol. There is the Windmill Hill City Farm and the Urban Centre for Appropriate Technology (the urban equivalent of the National Centre for Alternative Technology that has been in operation near Machynlleth for about ten years). Both these ventures are concerned with what individual householders and small neighbourhood groups can do for themselves in co-operation to meet some of their own basic needs and lessen their dependence on money income.

Totterdown is an inner city area, the heart of which was destroyed when 550 houses and a shopping centre were demolished to make way for a road that has not been built, leaving an area with few facilities divided by a wasteland. One of the few remaining retail businesses was forced to close as a result of the blight, and the premises have been let to Totterdown Shopping Community Action and Exhibition Centre Ltd at a peppercorn rent. Since 1981 most traders and workers in the building work closely together and meet weekly. They have formed the Totterdown Centre Co-operative with a view to purchasing the building and running it as a co-operative. There are retailing activities – shoes, fruit and vegetables, fancy goods, pet foods, antiques, haberdashery, wholefoods, and books. There is a wholesale bakery and a cafe. There are carpentry workshops and spaces for pottery, painting, silk screen printing, a darkroom, band practice facilities, and a bicycle repair workshop. Charitable organizations have stalls – Oxfam, Tools for Self Reliance, St Peters Hospice, etc. – and social activities for all ages, including a Children's Community Workshop.

Bristol Youth Workshops was set up in 1975 by a number of people who were concerned about the increasing rate of unemployment in the Barton Hill area of Bristol. It was one of many similar initiatives which developed in association with Manpower Services Commission schemes – the Job Creation Programme, the Youth Opportunities Programme.

New hope in shared workshops

The New Work Trust Company Ltd is a later development which emerged from consultations between various public and private bodies. The plan involved bringing into a working consortium a wide range of institutions and organizations to develop practical forms of assistance to new starters and existing small businesses. The initial aims included a workshop centre; business services; and training facilities. The Company is limited by guarantee and members may not contribute more than £100 each. There is support from some 250 different bodies – public and private – who together provide additional financial guarantees of £268,000. Between September 1981 and February 1982 Avondale Workshops were established. The 28,500 sq.ft of floor space was converted into 66 units ranging from 100 to 600 sq.ft each. Within eighteen months there was about 90% occupation, with over 200 new jobs created. Of the 73 ventures involved in its short life there have only been three liquidations. The success of the Avondale Workshops encouraged another conversion of some 60,000 sq.ft into 80 units at Station Road in the Kingswood district in 1983 and 23 units were quickly let. In addition to work-space accommodation the company operates a small firms marketing centre, a vocational trade skill training scheme, training in information technology, and business and management services. Enquiries for premises are running at a rate of 400 per annum, with as many small firms seeking assistance from the marketing service.

Aid to Bristol Enterprises, subsequently redeveloped as Bristol and Avon Enterprise Agency, is sponsored by seventeen leading companies in the area, the high street banks, the Bristol Chamber of Commerce, Industry and Shipping, and the Bristol City Corporation. The Agency offers advice and help to small firms through its own staff and also from the expertise of its sponsoring organizations. This covers a wide range of technologies, production engineering, purchasing, marketing, finance, computers, factory and warehouse planning, and industrial relations.

New hope through Local Enterprise Trusts

The commonest type of local economic initiative in both town and country areas is the work of Local Enterprise Trusts, alternatively referred to as Local Enterprise Agencies. Following a weekend meeting in September 1977 at Swindon, Wiltshire, sponsored jointly by the Intermediate Technology Development Group and the Foundation for Alternatives, it was agreed that Local Enterprise Trusts were likely to be the most effective way of mobilizing the fragmented and underutilized resources of a community in support of its own economic development. A Local Enterprise Trust was defined as 'a broadly based local group dedicated to the creation of worthwhile work through the fostering of small scale enterprises. It functions primarily by helping with access to technical and professional assistance/information, and by promoting collaborative arrangements as in, for example, marketing and the provision of suitable premises. It is a non-profit-distributing organization, which may (or may not) establish trading subsidiaries in pursuit of its primary ends.' During the following months a project to promote the formation and monitor the progress of six prototype LETs was established by John Davis of ITDG and Stan Windass of the Foundation for Alternatives. With assistance from Shell UK Ltd, BICC Ltd, and the Anglo German Foundation, a seminar, attended by 17 major companies and chaired by Sir David Barran, described the six prototypes – Antur Aelhaearn, West Somerset Small Industries Group, Hackney Business Promotion Centre, Community of St Helens Trust, Runcorn Business Link, and Clyde Workshops – and launched the project, which was funded by a grant made by the Gatsby Charitable Foundation to the Intermediate Technology Development Group.

Little more than a year later, Clive Woodcock wrote in the *Guardian*: 'The creation of 700 new jobs in one town in the space of twelve months through the encouragement of small firms is an achievement that would be the envy of anyone struggling with the unemployment problems arising from the changing patterns of industry'. The achievement he referred to was that of the St Helens 'local enterprise trust', led by its remarkable director Bill Humphrey. Under his careful guidance this initial success was maintained and developed.

The concept of a 'local enterprise trust' emerged out of a recognition that there existed in Britain no mechanism to help people trying to take the first steps, against enormous handicaps, in establishing viable wealth and employment creating activity. The situation was the same as it had been for hundreds of years, except that the path of the intending entrepreneur had been strewn with a multitude of different obstacles. Anyone who tackled this obstacle course had to be slightly mad or desperate; and anyone who managed to reach the conclusion, having set up a new venture, was heroic. Initially the promoters of the concept found themselves confronted by a catalogue of reasons why such an institution was an impossibility. Many reasons were given as to why entrepreneurs would not have any truck with any kind of aid mechanism, and as many more told of the absurdity of the notion of people volunteering unpaid assistance to small businesses whose purpose was to make money.

There was no other way of discrediting this mythology than putting it to a practical test – hence the six pioneering ventures launched in 1978. There were sound reasons for thinking that if entrepreneurs wanted advice and help, and if skilled people would volunteer to give it, there would be considerable benefit to the development of the small firm sector. Such valuable help was generally available to subsidiaries of large companies and was greatly to their advantage. Careful pre-startup work minimized the danger of launching a potentially disastrous new venture; and expert support during the first few vulnerable years of a new firm's life greatly reduced the high risk of premature failure. The creation of a 'local enterprise trust' could to a considerable extent provide the same kind of help to independent and free-standing small firms. By bringing all the many different kinds of assistance required into a single institution the entrepreneur would avoid the frustrations of going the rounds of different consultants. A 'one-stop-centre' was what the small firm needed, and one untainted by bureaucracy.

The resources required to operate an effective service are considerable. It was the firm conviction that there is within almost all British communities a great store of under-utilized resources, and the goodwill to make them available to a non-profit 'local enterprise trust', that made it possible to provide that service with very little expenditure of money. Cash was scarce, but kind was plentiful and available. It was found in both the public and the private local institutions and organizations – in the local authority, in educational establishments, in local businesses and factories, in banks, trade unions, Chambers of Commerce, Trades Councils, and among individual citizens both young and old, male and female. Inevitably some of the bigger organizations were able to offer considerable support to the Trust. As a matter of firm principle it was determined that it was essential that the Trust should not only be absolutely independent of any dominant local forces, but it should be seen to be independent. For this reason it was particularly important that the full-time director of the Trust should be an independent person responsible only to the Trust.

Another important principle was that it was people with problems or plans that were to be served; and the temptation for the Trust to dream up business schemes and try to fit people to them had to be resisted.

In the press release announcing the establishment of the Community of St Helens Trust the aim was clearly defined as 'to increase prosperity and improve employment opportunities through the encouragement of new business enterprises'. It stated that it was to achieve this aim by the community marshalling its many resources through a non-profit-making Trust, whose Trustees would represent a wide cross-section of the St Helens community. From the start it was made clear that it was not concerned with lending financial or other support to people with schemes that could not become self-sustaining. Apart from a small emergency reserve fund the Trust had no resources of its own and was dependent entirely on those of its supporters. In soliciting support it appealed as much to enlightened self-interest as it did to social conscience.

After some two years of operation the director listed a dozen lessons learned. Two of the most important ones were that scarce finance is not the serious problem that it is often assumed to be, but premises are a major difficulty. Skill in gaining access to available sources of finance is the key to resolving most money problems in a business start-up. Conversion of existing disused premises into multi-occupation small work-space units on short-term licences is often the most cost effective way of meeting start-up needs. The St Helens Trust promoted such conversions. With their advice and assistance an 84,000 sq.ft steel fabrication factory was converted imaginatively by Arthur Cox into units ready for letting, according to the Trust's suggestion of waiting for clients to make known their requirements. Explaining how he operated, Cox said: 'People simply mark out on the floor the area they require and we build a wall to divide them from the next business.' In another local enterprise trust in West Somerset the same principle was applied by Fred Wedlake, but in this case the people themselves built the partition walls and carried out the fitting and decoration. In this way a sense of individual pride of ownership developed which has continued with a high standard of maintenance and an absence of vandalism.

Not all the work of an enterprise trust has such visible manifestations. Much of it concerns bringing influence to bear and applying pressure that breaks log jams. Equally important is the unromantic role of simply being available to business men and women to talk over their problems freely and in complete confidence and without obligation. Small business management is a very lonely occupation and a trusted friend is invaluable.

The St Helens Trust has a special place in the development of the

Local Enterprise Trust (LET) movement. Not only has it been consistently successful in the pursuit of its aim to create prosperity and employment opportunities, it has promoted two important new local initiatives and a national one. A community approach to the important problem of work training and a local venture capital company would most likely not have occurred had it not been for experience gained from the Trust operations. At national level the importance attached to this approach by Sir Alastair Pilkington and his colleagues, which was forcefully expressed in his Chairman's address at the 1979 Annual General Meeting of Pilkington Brothers Ltd, was an important factor in the development of the 'Business in the Community' organization in 1981. It has gained wide support from major industrial and commercial companies, financial institutions, the TUC and various voluntary bodies. Largely as a result of its work the number of local enterprise trusts or agencies throughout the United Kingdom exceeded two hundred in 1985. That has only been possible because of the proven effectiveness of the St Helens example and that of the other pioneering ventures.

One of the commonest questions asked about Local Enterprise Trusts is 'What kind of firms are helped to start up?' It is a difficult question to answer exhaustively because the range of activities is diverse: to most enquirers the most surprising thing is that a high proportion are in some kind of manufacturing.

After the Clyde Workshops had been in operation for a year, accommodating more than 50 small firms, the following activities were taking place: precision engineering, furniture manufacture, printing, electronics, food processing, shopfitting, cable drums, doll making, metallization, welding, light engineering, plumbing, joinery, oil rig servicing, pottery, central heating, catering, television repair, ventilation, brickmaking, electrical engineering, clockmaking, coach building, chemicals, radio car service, builders, electricians, roofing contractors, wrought iron fabricators, pallets, sheet metal, armature winding, vehicle repair, cooperage, spray painting, timber crates, scientific instruments, refrigeration, draughtsmen, instrument mechanics.

Another common question is how much employment is created by LETs and what does it cost? The truthful answer is that LETs themselves create no employment: the entrepreneurs produce it all. Estimates have, however, been made about the numbers created with the assistance of a LET, or saved in an existing company as a result of help received. They vary from place to place and they obviously depend upon the skill, competence, and commitment of the people running and serving them – and no two people are the same. A good LET can be associated with about 200 new jobs, or jobs saved, each year. Since for every wealth-creating job produced another is generated indirectly, a total of 400 per annum is roughly the upper limit, and average performance is about half of that – 200 direct and indirect jobs each year. With more than 200 LETs in operation in 1985 they could altogether be concerned with 40,000 jobs in that year; and between then and 1995 with nearly half a million.

The true cost of a LET operation cannot easily be expressed in monetary terms because most of the work is given in kind, through voluntary skill and time and in the form of secondments from local organizations. The monetary cost (mainly the salary of the independent director and his overhead costs) is usually between £15,000 and £30,000 per annum. Therefore, in monetary terms alone, only some £300 per direct job is involved. This is an almost insignificant amount for a community to invest for employment creation. The pay-off in terms of value added, much of which is spent in the community, lies between £7,500 and £10,000 per annum for each job; for a town of 50,000 population this will increase prosperity (net income in the community) by between three-quarters of a million and two million pounds each year. Considering the vast Exchequer grants doled out to attract industrial investment to a locality for the addition of small numbers of jobs, the LET route is very cost-effective.

Apart from increasing the population coverage of the UK with this mechanism of generating prosperity and employment that is widely distributed through many small firms, what other potential exists for Local Enterprise Trusts? St Helens has demonstrated that there is an advantage in developing local schemes for training and retraining to match the needs of the employers and employees in the community. There is growing evidence of a serious mismatch as demand grows for new skills all over the country. With the Rainford Venture Capital initiative, St Helens has demonstrated the feasibility of a fresh approach to risk investment which is badly needed for certain types of new business. It is common for much industrial and commercial property to be owned by absentee landlords.

Local Enterprise Trusts have demonstrated their capability as developers of property for small businesses. They would appear to be a useful vehicle for sponsoring schemes that could recover the ownership of much local property for the community's benefit as well as an object for local savings.

Many LETs have provided low cost routine services for the small firms in their locality. In most parts of Britain domestic maintenance and repair services are in a primitive state, with erratic quality and poor reliability. Local Enterprise Trusts could arrange the creation of a community business or businesses which would organize services of a high standard of quality and reliability on a scale of charges closely related to actual labour and material basic costs. Such an arrangement would permit selected, well qualified self-employed craftsmen to operate independently without the necessity of taking on the running of a one man business. It would also have an impact on 'cowboy' operations.

Some small businesses are capable of replication by means of franchising. LETs could develop subsidiary franchise companies whenever the appropriate kind of business appeared in their locality. As subsidiaries of a non-profit-making LET, franchisees would be assured that the terms on which their licence was run left most of the benefits with those responsible for the success of the franchise business.

Although there is no evidence that the supply of potential entrepreneurs is being exhausted, there is nevertheless an immediate need for local schools and colleges of further education to increase interest and involvement in small enterprises and to help create a positive spirit of initiative, entrepreneurship, and achievement.

It would be wrong to leave readers with the impression that small firms or Local Enterprise Trusts are the answer, or even a major part of the answer, to the current high level of unemployment. Despite the success of the St Helens Trust the level of unemployment in the town was higher in 1984 than it was in 1978. Redundancies in traditional employment have taken place faster than the creation of new job opportunities. However, that situation could improve over the next ten years now that serious overmanning has largely been eliminated and the least competitive firms have collapsed. There can be no doubt that Local Enterprise Trusts can make an invaluable contribution to the creation of new wealth creating employment on a significant scale, both by increasing the rate of new firm formation and by reducing the death rate of established firms. Last but not least they can help to forestall domestic disaster and ruin. In the words of Chris Barnes, the manager of the Lowestoft Enterprise Trust, they can head off people having attitudes such as:

'I know I've only just started up, but I need a business loan to go on holiday.'
'I couldn't miss Saturday afternoon football to go on a business course.'
or
'It's easy. It must be. The guy I worked for was making a fortune.'

One thing local initiatives do not do is reduce the degree of commitment that is called for from everyone who sets up and runs a small business.

CHAPTER 10
Small beginnings

No land is an island

This book focuses on the British economy. We have compared Britain with other industrial nations and noted many ways in which Britain differs from them. In identifying a need to redress the balance between small and big business, combined with a greater emphasis on repair, reconditioning, re-use and recycling, we have seen that the problem is greater and more urgent in Britain than in other countries. Nevertheless the same basic change of direction from a 'produce, consume, and throw away' system to a conserving system is required of all the advanced industrial countries, and the economies of the Third World countries will also need to develop on a 'conserving' model if they are not to find themselves impoverished in a world which makes them the victims of increasing real prices as material resources become scarce.

Resource scarcity in a world of expanding population is not the only reason why there needs to be a global shift to 'conserving' economies. There is a growing pile of studies of the catastrophic effects which continuing development on traditional lines will inevitably have on the natural environment within the next few decades. The Global 2000 Report, prepared for the President by the US Department of State and the Council for Environmental Policy in 1980, painted a grim picture of what will happen by the year 2000 in the absence of radical changes in the way that the economically developed world behaves in respect of consumption. The bleak prospect is illustrated by four points:

1. A 90% increase in world food production would still leave per capita consumption in South Asia, the Middle East, and the less developed countries of Africa either with little improvement, or in some cases a serious decline below present inadequate levels.
2. For the quarter of the human race which depends entirely on wood for fuel, a devastating 25% gap between requirements and supplies will develop.
3. By the year 2000 water supplies in many developing countries will become increasingly erratic as a result of extensive deforestation. Of the remaining forests 40% will have gone by the end of the century.

4. An area the size of the State of Maine is becoming barren wasteland every year; and as many as 20% of all known plant and animal species could be lost forever in the next two decades.

Trends in both the quantity and the nature of employment, as technology becomes more complex and sophisticated, is creating a serious division in society, in which many jobs are available only to a small elite of workers and the mass of the population are increasingly disqualified from employment. Another adverse effect of increasing capital intensity and the concentration of businesses into a smaller number of very big companies is the individual's loss of opportunity to participate in matters that affect their own lives and communities. The absentee landlord has reappeared in many new forms; the absentees become uncontrollable and as oppressive and threatening as the original rural form.

The need to conserve scarce resources, to halt the catastrophic destruction of the natural environment, and to avoid the emergence of despotism and dehumanization is not confined to Britain: it is of worldwide importance. Nevertheless, as the first industrial nation, and the one that has gone furthest down the road of centralization and concentration of economic activity, Britain has a special obligation to be in the vanguard of progress in a new direction of sustainable development.

A future for an industrial nation

In previous chapters we have outlined and illustrated such changes as can be made, even under the adverse rules, regulations, and practices that exist to promote and sustain the 'consumerist' society. A great deal of production work can be carried out efficiently in small units and in small or medium-sized firms, leaving a small proportion – mainly primary materials production industries – in the big firm sector. There is plenty of evidence that much production can be carried out satisfactorily at modest levels of capital investment, avoiding the complexity and heavy overhead costs of many of the present companies. There is no net gain if the economies of scale of costly capital production plant are lost through the diseconomies of big, complex business systems and the increasingly costly distribution of products. Once it is recognized that labour-intensive repair and reconditioning work can be in many instances more effective in terms of 'adding value' than the equivalent capital-intensive manufacturing operation, a whole new world of satisfying, skilled, widely dispersed, small-unit employment can be promoted. Once these lessons have been fully understood, a major shift in technological development can take place which will be geared to meet the needs of the new generation of small firms.

Progress will, however, be limited so long as the existing rules controlling economic and social behaviour are retained. A new set of 'rules of the game' needs to be developed. This is not something that can be well done in a planning office. The changes that need to be made should be a response to actual experiences, a response that will depend on feedback from the grass roots to legislators. That will require the existence of intermediary bodies that can represent the experiences of the new economic frontier. The Local Enterprise Trust movement can play an important part in such a process of communication.

The Golden Rules of Appropriate Technology

To be coherent the new set of rules needs to be guided by some general principles. The same principles will be those which influence all those who play a part in changing the direction of economic development. They may be described as the 'Golden Rules of Appropriate Technology'. Three such principles can be expressed as:

Rule 1 Minimize the consumption of non-renewable natural resources by giving priority to activities which maintain, repair, recondition, re-use, and recycle durable goods, property, and materials (the 4Rs).

Rule 2 Operate on as small a scale as possible consistent with efficient use of resources.

Rule 3 Operate with the minimum amount of capital required to provide the rate of 'adding value' necessary for the prosperity of the total enterprise system.

Adherence to each of these principles contributes to most, if not all, of the following objectives.

1. Minimize the consumption of non-renewable natural resources.
2. Reduce environmental pollution and waste.
3. Sustain a healthy natural environment.
4. Provide a wide range of satisfying employment.
5. Provide a mix of employment that matches the abilities of the whole working population.
6. Provide employment that is capable of a level of 'adding value' sufficient to ensure private and community prosperity.
7. Ensure that wealth creation is geographically as widely spread as is possible.
8. Minimize the amount of 'added value' that goes to the suppliers of capital, and maximize the proportion taken by people involved in its generation.

9.   Provide a greater degree of local economic independence and internal interdependence.
10. Minimize problems of distribution.
11. Distribute power and responsibility as widely as possible.
12. Reinforce the cohesion and solidarity of families and local communities.
13. Facilitate international convergence of economic development.

These Golden Rules represent a complete reversal of the principles that have guided development philosophy since the Second World War. It was only when the large-scale philosophy of development was pressed to the limit, indiscriminately applying the idea that 'bigger is always better', that most of the damage was done. In applying the three Golden Rules great care must be taken to ensure that we do not fall into a similar error of thinking that everything, regardless of the nature of the activity, must be small, and that everything would be better if it were smaller. There is, as it were, an ecology of economic activity: each category of activity has an appropriate size and requires 'appropriate technology' with which to operate.

This radically different approach to development is ultimately a condition for peace and justice. As Schumacher once said: 'The real pessimists are those who declare it impossible even to make a start.'

Appendix

Things for people to do

Although it is evident what, in general, needs to be done in order to effect the change in direction of development that is advocated in this book, many readers will wish to consider what opportunities exist for a personal contribution to that end. It is certainly not the intention of the authors to tell people what they ought to do; nevertheless it could be helpful to list some of the many opportunities that exist for individuals in their different roles as consumers, as members of local communities, as members of organizations and as either employees or employers.

As consumers – but without being excessively scrupulous!

1. Adjust your diet so that you eat healthily, using as much fresh food – locally grown where possible – with a seasonally variable content.
2. Make your own preserves to replace commercially processed products. (This saves a great deal of packaging materials.)
3. Make as much processed food and drink as possible at home.
4. Cook with gas, an Aga, or a microwave.
5. Purchase as much as possible from local small shops.
6. When shopping provide your own re-usable bags or containers.
7. Purchase loose products rather than pre-packed.
8. Insulate your home.
9. Avoid electric heating, if possible.
10. Use showers in preference to baths; and use cold water in summer: it is very refreshing!
11. In winter rely as much as possible on clothing to keep you warm in preference to space heating.
12. Use minimum background space heating, topped up as necessary with spot heating.
13. Walk as much as possible; cycle or use public transport when convenient in preference to a car.
14. Consider taxis, car sharing, or car hire as an alternative to car ownership – particularly if your annual mileage is low.
15. Aim steadily to increase the proportion of your spending on 'services' rather than on 'goods'; and offer work to local tradesmen.

16. Purchase 'used' goods in preference to 'new' goods.
17. Try to ensure that when disposing of something it can be reused.
18. Try to ensure that some of your savings are invested in small enterprises.
19. Prolong the life of articles by proper care; and avoid throwing away articles if they can be repaired.
20. Progressively increase your giving.

As members of local communities

1. Make sure that land is available for allotments.
2. Organize community cold storage in preference to home freezers.
3. Ensure that there are proper provisions for recycling domestic waste (bottle banks, paper collections, etc.).
4. Ensure that there are adequate 'swap-shop' schemes.
5. Ensure that there is proper provision for cycling and public transport.
6. Ensure that there are skilled maintenance and repair services, with proper training facilities and discouragement of 'cowboys'.
7. Ensure that there is suitable commercial property, with rents and rates to suit small firms.
8. Ensure that as much as possible of commercial property is owned and controlled by the community in the interests of the community.
9. Ensure that there is emphasis in the upbringing of young people on a 'conserver' style of life.
10. Ensure that there are local organizations which promote small scale enterprise and resource economy, and serve their needs (e.g. Local Enterprise Trusts and Local Energy Groups).
11. Promote the means whereby local finance can be invested in local economic activity.

As members of organizations

1. Avoid actions that could be counter-productive to the development of a 'conserver' economy.
2. Promote lobbying of local and national government to help bring about legislative and regulatory changes that are needed for the full potential of a 'conserver' approach to be realized.
3. Encourage the development and acceptance of standards and codes of practice that will restrain the emergence of an exploitative and poor quality small business sector, and will ensure that a high level of quality is attained.

4. Ensure that there is a transfer of public resources in favour of small scale, resource economical operations.
5. Seek to influence public opinion and political party policies in favour of a 'conserver' economy.
6. Encourage the organization to trade with small firms and adopt resource economical practices.

Lightning Source UK Ltd.
Milton Keynes UK
UKOW030821210613

212605UK00005B/57/P